TRANSACTIONS
of the
AMERICAN PHILOSOPHICAL SOCIETY
Held at Philadelphia
For Promoting Useful Knowledge
Volume 93   Pt. 2

THE 18TH CENTURY CLIMATE OF JAMAICA

Derived from the
JOURNALS OF THOMAS THISTLEWOOD, 1750–1786

Michael Chenoweth

AMERICAN PHILOSOPHICAL SOCIETY
Philadelphia • 2003

ISBN: 0-87169-932-X
US ISSN: 0065-9746

Library of Congress Cataloguing-in-Publication Data

Chenoweth, Michael, 1960-
    The eighteenth-century climate of Jamaica derived from the journals of
  Thomas Thistlewood, 1750-1786 / Michael Chenoweth.
        p. cm. — (Transactions of the American Philosophical Society,
      ISSN 0065-9746; v. 93, pt. 2)
    Includes bibliographical references and index.
    ISBN 0-87169-932-X (pbk.)
      1. Jamaica—Climate—History—18th century—Observations. 2. Climatic
  changes—Jamaica. I. Title: 18th-century climate of Jamaica derived from
  the journals of Thomas Thistlewood, 1750-1786. II. Thistlewood, Thomas,
  1721–1786. III. Title. IV. Series.

  QC987.J3C48 2003
  551.697292'09'033—dc21
                                                                    2003048158

Design and typesetting
Maria Karkucinski
Book Design Studio

THE 18TH CENTURY CLIMATE OF JAMAICA

CONTENTS

chapter **6**

LIST OF TABLES

LIST OF FIGURES

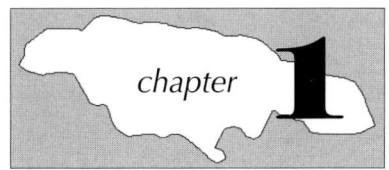

chapter 1

THE UNKNOWN CLIMATE HISTORY OF THE CARIBBEAN

Modern research has revealed that much of the Earth was locked in a much colder climate prior to 1850 (Bradley and Jones 1992; Lamb 1977). The coldest phases varied from one region of the world to another. Lamb (1977) considered 1550–1700 to mark the major phase common to most regions and the period 1780–1850 to be another important epoch of generally colder global temperatures. At times, the extremes of climate became so marked that there was public speculation as to the causes (e.g., the coldness of the year 1789 was seen to require "meteorological investigation" by a correspondent to the *London Times* of 14 July 1789). Another observer speculated that the spate of violent hurricanes in tropical regions of the Atlantic Ocean was related to the vehemence of storms in the temperate zones to the north at almost the same time (*London Times*, 20 November 1786).

The climate prevailing away from western Europe and the northeast fringe of North America in the eighteenth century is poorly documented. Elsewhere in the world, except for scattered British colonial records from India,[1] the documentary evidence elsewhere in the world is either unavailable, difficult to access, or has been unknown to climate historians. The tropical regions of the Americas are almost completely devoid of daily weather records prior to 1800 and the few published records are usually very short summaries (Byam 1755).

Climate in the Tropics affects the great majority of the human race because most of the world's population is concentrated in tropical and subtropical regions. Temperature ranges, although smaller than in colder latitudes, vary over large regions in

---

[1] Most of these records are in the Meteorological Archives of the Royal Society, London.

coherent patterns reflecting tropical oceanic temperature fluctuations in the Pacific Ocean (El Niño and La Niña events). When the Trade Winds weaken over the eastern tropical Pacific Ocean, the normally cold upwelling currents that cool the west coast of South America are replaced by unusually warm surface waters. In the El Niño phase, the main centers of rainfall formation move away from their normal regions (Indonesia) to areas that are normally arid (central near-equatorial Pacific islands and the deserts of Ecuador and northern Peru). Patterns of rainfall vary dramatically throughout the global Tropics and in some regions in higher latitudes.[2]

El Niño and La Niña play an important role in the climate of the Caribbean region and Central and South America. Caribbean sea and air temperatures lag by several months, but vary in phase with, the warm El Niño and cool La Niña phases of the tropical Pacific Ocean. Rainfall patterns are more variable but in the winter months toward the conclusion of an El Niño tend to be unusually wet in Mexico and the Gulf of Mexico and unusually dry along the northern coast of South America (Ropelewski and Halpert 1987; Halpert and Ropelewski 1992). Hurricane formation in the tropical Atlantic Ocean south of about latitude 25°N is inhibited by the upper level wind patterns produced in warm El Niño events. The opposite pattern is produced during the cool La Niña events (i.e., enhanced numbers of hurricanes forming south of about latitude 25°N) (Landsea et al. 1999). Although locally very destructive, hurricanes and tropical storms provide a substantial proportion of a given year's rainfall to those locations they most frequently visit (Cry 1967).

Long-term variability in the number and intensity of El Niño and La Niña events is a subject of great interest to climatologists. For historians, an improved knowledge of such events will provide a better understanding of how natural climate variability worked with man-made events and causes to affect historical events. Should real and persistent anomalies in climate be found, they may useful in assessing their role, if any, in historical processes on a range of time scales.

Virtually nothing reliable can be said about "average" temperature and rainfall in the Tropics before the nineteenth century (and even the nineteenth century record has significant temporal and spatial gaps). To date, no reliable temperature record from the Tropics is available from the eighteenth century to compare with

[2] An online tutorial, with excellent graphical visualization tools, can be found at the Web site *http://www.cpc.ncep.noaa.gov/*.

modern data. Research on corals growing in the waters near Puerto Rico indicates significantly colder temperatures in the 1780s (Winter et al. 2000). Some rainfall measurements were made in the eighteenth century in the tropical Atlantic (Byam 1755; Long 1774), but no individual long time series has previously been reported.

The climate of the Caribbean is also influenced by the position and the strength of the semipermanent area of high pressure over the eastern Atlantic, known as the "Azores High" and in the United States (less accurately) as the "Bermuda High". Changes in the position and strength of the Azores High occur annually, as part of the normal seasonal cycle of the weather. During the winter months, it is further south and covers a less extensive area. During the summer, its center moves north to a position slightly to the southwest of the Azores, and it expands to cover a much larger area of the Atlantic Ocean. Over longer periods of time, the position and strength of the Azores High can persist in unusual locations, with dramatic implications for the Caribbean climate.

When the Azores High is stronger than average, it produces an increase in the strength of the Trade Winds that blow along its southern side in the Tropics. These stronger winds produce larger seas, and mix the surface layers of the ocean more thoroughly. This increases evaporation of seawater into the air and lowers the air and sea surface temperature. One result is that the atmosphere becomes more stable, and less prone to produce rainfall. Tropical waves and tropical lows are weaker than normal, and produce less rainfall than average.

When the Azores High is weaker than average, the Trade Winds are reduced in strength. The result is a lowering of the evaporation rate from the ocean to the air, warmer than usual sea and air temperatures, and an unstable atmosphere that will produce more and better organized areas of rainfall and storms. The topographically flat islands of the Tropics remain at the mercy of individual showers and storms, but there are more opportunities for rainfall when the Trade Winds weaken (all other things being equal). Islands with large mountain chains are less vulnerable to water shortages because the forced ascension of the Trade Winds over them forces the air up and cools it, producing clouds and eventually rainfall. The exact distribution of rainfall on these islands is highly dependent upon the geography and the prevailing winds during the rain events.

The long-term record of Trade Winds over the world's oceans over the past 500 years is virtually unknown before 1850. While ships' logbooks and scattered other records are available, the massive amount of data contained in these records are not available to climate researchers. The only digitized collections of tropical weather data from ships previous to 1850 are a collection of about 65,000 noontime weather reports made by the English East India Company, the U.K. Royal Navy, and other ships, mainly from 1807–1827 (Chenoweth 2000), and the Maury Collection data, mainly from the mid-1830s to mid-1850s (U.S. Department of Commerce 1998). Very few of the Atlantic Ocean observations are from the Caribbean region.

The modern record of climate reveals relationships between climate in the global Tropics and some extratropical areas ("teleconnections") (Hoerling and Kumar 2000). The stability and dependability of the teleconnections may or may not be true before the twentieth century. Historical records will allow a much longer time period to be dated exactly and the patterns can be studied for any long-term change. Knowing the amount of past variability will allow computer modelers of future climate to better interpret their forecasts based on past behavior. In turn, this influences the assessment of the amount and significance of any forecast climatic change for a given region. Because these climate forecasts are helping to drive nation-state and international governmental agency responses and nonresponses to global climate change, they require the best historical information that the available data can provide. Ultimately, an improved knowledge of past climatic variability will assist us in determining if the recent changes in climate are attributable to the long-term increase of so-called "greenhouse gases" and other man-made pollutants in the atmosphere.

EIGHTEENTH CENTURY OBSERVATIONS AND THEORIES OF CLIMATE

Fleming (1990) provides a concise and comprehensive survey of the issues concerning weather and climate that were foremost among scientists in the eighteenth and early nineteenth centuries. The great extremes of climate in North America, compared with that of Europe, were unexpected given the expectation based merely on the latitude of newly colonized lands. The clearing of forests in the New World was widely thought to be likely to change the climate by changing both precipitation and temperature patterns. Whether the climate was becoming colder or warmer, drier or wetter was open to debate.

Thomas Jefferson, writing sometime before the mid-1780s, summarized a view that Colonial America's climate was becoming less harsh than in earlier years. "A change in our climate, however, is taking place very sensibly. Both heats and colds are become much more moderate within the memory even of the middle-aged. Snows are less frequent and deep. . . ." Other observers, usually in different places and in other years, came to the opposite conclusion (Fleming 1990).

The situation was no easier for Europeans to decide. The weather in early 1786 across the United Kingdom is succinctly condensed in *Palmer's Index to "The Times" Newspaper* for 1786 by date, page, and column number. Under the heading "Weather" we read (partially excerpted):

> almanack makers' prediction of mild weather proves false, JAN. 3, 3a; Serpentine and Thames frozen, 3 3c; severe frost hampers work in fire fighting, 3 3c; heavy snow blocks roads in north-east, 6 3c; thaw following heavy snow makes travel difficult, 9, 2b; coldest weather remembered since 1740 in Glasgow threatens business and cause death of chaise driver, 12, 3d; thunder, lightning, frost and snow occur simultaneously on south coast, 16 2b; snow blocks roads again but likely to be cleared by rain, 17, 2c; man saves woman almost buried in snow between Coleby and Lincoln, 17, 3d; heavy snowfall blocks roads in south-east, MAR. 4 3d; heavy snow and severe frost reported in neighbourhood of York, 7, 3d; snow delays mail coaches for Bath, 7, 3d; snow reported seven feet deep in neighbourhood of Turnbridge, 13, 3c; nearly 100 men employed in clearing road after heavy fall of snow, APR. 1, 3d . . .

Four years later, in early 1790, the same newspaper provided evidence of anomalous winter warmth.

> WEATHER mildness contrasted with frost in Rome, Jan. 4, 3a; unusual lack of snow and ice and abundance of wind and rain in Cumberland and Westmorland, 13, 3b; fields of rape flower at Holderness; gooseberries gathered at Whitehaven, 22, 3b; mildness of season reflected in blackbird's nest with four eggs at Leyton, 27, 3a; forwardness of season further illustrated by swarming of hive bees at Cambridge, MAR. 2, 3b; temperature at Edinburgh on average eight and half degrees Fahrenheit higher in March 1790 than in March 1789, APR. 8, 3a . . .

The big differences in weather can be attributed to normal year-to-year variability, so it is not surprising that the theories of climatic change that were debated in the eighteenth century were never resolved. Even today, with overwhelming evidence of global climatic change, there are differences of opinion as to the scope and significance of changing climate. In the eighteenth century, even if the climate was not changing, it was subject to fluctuations of various lengths and degrees of severity. Without central heating, air conditioning, refrigeration, and other amenities that today remove people further from the natural world's direct and immediate influence, the weather and climate of a location placed important restrictions on the comfort, health, food consumption, water quality and availability, and transportation of eighteenth century populations. Most people produced some or all of their food for local consumption. When crops failed, people went without or had to purchase foodstuffs from other areas. The weather was always a cause for concern but especially when it threatened incomes and lives.

In the Caribbean, sugar was the most important crop grown. The middle and late eighteenth century marked the heyday of sugar production and profits in the region. Sugar planters followed the annual cycle of climate in establishing the dates to plant, cultivate, harvest, and export their crops to Europe and North America. Canes were normally planted in the months of October through December, when the moist ground could be easily worked and the canes established before the dry season began. During the months of January through May, the canes previously planted (at slightly lagged intervals to account for the fourteen to sixteen months needed to reach maturity) were harvested and processed in the sugar mills. Slave labor was used for most of this work. During the rainy season, from June through September, there was relatively little activity other than routine weeding and hoeing of the cane crops. However, the rains that fell helped to determine the ultimate condition of the next season's crops. Too little rain could result in reduced growth and yield; too much rain could produce flooding that could wash away crops. Tropical storms and hurricanes were also feared, because the strong winds could completely break the canes and destroy windmills, sugar boilers, warehouses, and other properties needed for the large-scale production of sugar. One major hurricane could destroy an entire year's crop, destroy all previous investments in equipment, and kill slaves (the most valuable investment, in the eyes of the slave owners) (Dunn 1972).

Jamaica became the largest sugar producer in the English-speaking Caribbean in the eighteenth century. By the 1780s, five-eighths of the crop received in England came from Jamaica (*London Times*, 18 August 1788). The crop harvested in the spring of 1788 was expected to exceed that of any other year for the previous fifty years and to be as good in quality as in quantity (*London Times*, 30 April 1788). Good crops were also reported in 1789 and

1790 (*London Times*, 8 January and 28 April1789; 10 June and 13 August 1790). On the eve of the Napoleonic Wars, Jamaican sugar planters had a bright future.

Near the start of this most prosperous period for Jamaican sugar planters, in May 1750, Thomas Thistlewood arrived from England to make a new life. With letters of introduction, he sailed to Savanna-la-Mar, in western Jamaica, to look for work as a plantation overseer. Thistlewood had surveying and mathematical skills to offer, was keenly observant of his new surroundings, and in his late twenties was eager to finally settle down to make a proper living. Thistlewood never became a sugar planter and was certainly not rich, but he did prosper. So much so that he was able to purchase a small plantation and thirty slaves. He ran a profitable estate by hiring out his slaves to bigger planters in the region, selling his services as surveyor for the numerous undeveloped properties in the area, and growing and selling rare and other specialty plants and flowers from his garden. Thistlewood maintained one of the most extensive horticultural practices kept on the island of Jamaica in the eighteenth century. His estate had many well-known visitors from England and from throughout the island of Jamaica. His garden and his scientific knowledge gave him an entry into the leading families of Westmoreland Parish and a reputation throughout the island (Hall 1989). This was recognized when his obituary was written up and published in a Montego Bay newspaper, *The Cornwall Chronicle*.

> *Deaths. . . In Westmoreland, on Thursday the 30th of November, in the 65th year of his age, Thomas Thistlewood, Esq., a gentleman whose social qualities, during a residence of upwards of 30 years in that parish, had greatly endeared him to the whole circle of his neighbours and acquaintances, and whose attainments, in many branches of natural knowledge, in which he was peculiarly communicative, rendered him a most desirable companion to men of science.*[3]

Thistlewood never became a major player in Jamaican government or economics, but he was able to consider himself at least a part of the landed and educated establishment on the island. His garden and his weather were probably his great loves, the things to be most savored when the brutal work of slave ownership (which he took for granted, unthinkingly and at times with sadistic relish) and earning his living could be briefly set aside.

As will be shown in Chapter 3, Thistlewood was a remarkably keen weather observer who was years ahead of most of his contemporaries in his observing practices and habits. The cultivation

[3] *Supplement to The Cornwall Chronicle*, No. 704 [Supp. No. 625], Dec. 16, 1786.

of a successful garden required a close look at the weather. He placed great stock in his own power of observations, and knew what the leading advocates of weather observation methods were suggesting. He wrote his opinions on the possible changes in rainfall in Jamaica caused by the clearing of forests and (correctly) rejected the theory.

Although separated by 1400 miles and, no doubt, political differences, Thistlewood's slave ownership, his garden, his weather record, his critique of theories of climatic change, and his mathematical grounding in surveying reminds one of Thomas Jefferson. Both Jefferson and Thistlewood made weather observations over many years (Thistlewood did a better job), and cultivated their gardens with great care and industry. They would have had the potential for mutually enjoyable scientific discussions, had they ever met. Today, Monticello remains as a monument to its owner. Breadnut Island, the plantation Thistlewood purchased and developed, is no more. Jefferson's place in history is known and virtually every aspect of Jefferson's life, with the exception of his weather record, has been minutely studied. Thistlewood's daily journal provides a basic account and incredibly rich information for understanding slave society in Jamaica, but reveals little about the man. In the reverse of Jefferson, his daily weather record may help us better understand Thistlewood as a person and is the most important daily record of the earth's climate from anywhere in the Tropics in the eighteenth century yet available. Thistlewood's previously unrecognized legacy is in his weather record, and the questions both he and Jefferson pondered are today given a chance to be better answered by it. Their efforts were not pointless exercises in record keeping, but rich documentary evidence that allows global climate change of today to be better measured and understood.

chapter **2**

METHODS IN HISTORICAL CLIMATOLOGY

The usefulness and accuracy of Thistlewood's weather data belong to the field of historical climatology. The methods of historical climatology are addressed here in an overview that in no way touches on all the details of the methodology. However, as there is no general reference work available on this topic, it may serve as a brief introduction and give the reader an appreciation that climate reconstruction from historical documents can provide important information on the recent history of the Earth's climate. The general reader may, without serious loss, continue to Chapter 3.

Historical climatology is a subfield of climatology that focuses on written documentary sources in the era before the establishment of national and international organized weather-observing networks. Modern observing networks employ consistent standards, procedures, and instruments for measuring, recording, transmitting, exchanging, and archiving data. Meteorologists, climatologists, and other Earth Sciences professionals work with data derived from instrumental observing systems (satellites, ocean buoys, ship data, airfield weather stations, radar, automated weather stations, and the like) that conform to performance and reliability standards that are discrete and measurable. The margin of error in the data recorded from any given instrument is of interest since nonclimatic biases in weather data will provide inaccurate and misleading results. Research into global environmental change, and global warming in particular, needs data that are accurately derived, and from which uncertainties in the derived values can be quantified, to assist in the detection of potential and actual influences of anthropogenic effects on the Earth's climate. Even in modern observing networks, great care is required to remove these nonclimatic biases. For historical climatologists, the method is similar.

Modern weather-observing networks are designed for measuring the weather in the here and now. Climatologists work with these data to produce climatologies from networks that are not ideally suited for creating time series of weather variables that are free of nonclimatic biases. In the United States, the U.S. Army Signal Corps was the first government agency that was established and funded by an act of Congress to establish and maintain a national weather-observing network. Most of the weather stations were built along the nation's coastlines and inland lakes and waterways in urban environments. The instruments were usually placed on the roof of multi-story buildings in rented premises (Chenoweth 1993). The result was a weather network ideally suited for monitoring individual storms and providing a warning service to the public of approaching severe weather. What it was not suited for was providing the beginning of a continuous and homogeneous series of data for assessing changes in climate over many years.

Several factors limit a weather-observing network from being simultaneously a climate-observing network. First, the weather-observing network is not designed for monitoring on the inter-annual and longer time scales needed by climatologists. Second, the location of weather stations is constantly changing. In the United States, the establishment of airfields and the demands of aviation safety produced a requirement for airfield weather data and aviation-forecast services. The U.S. Weather Bureau responded by establishing new weather stations at these airfields (Chenoweth 1993) and, with time, closed most of the urban offices. Wind, temperature, and rainfall measurements once made fifty or one hundred feet or even higher, above the ground were moved to elevations between five and thirty feet above the ground. Third, urban heat-island effects produce pronounced effects on local weather that are not representative of the larger nonurbanized region. As cities grow and land use changes are made, the climate is artificially warmed (a nonclimatic bias) and the area affected by this artificial warming expands. Suburban growth quickly surrounded rural airfields, while manmade warming affects the temperature being recorded at these sites (Karl et al. 1986). Fourth, instrumentation and procedures change over time. For example, thermometer types include a mix of liquid-in-glass thermometers, thermographs using bimetallic strips with temperature-dependent coefficients of expansion, and remote-reading thermometers providing data electrically to a weather office tens or hundreds of feet distant. The days of a human observer walking outside to read a thermometer in a wooden louvered screen are decades past in modern weather-observing networks. Similar developments have affected all other weather instruments (Mitchell 1953).

Procedural changes involve the timing and frequency of observations and the exposure of instruments to the elements. In the late nineteenth century, instruments were normally read three times a day for use in nationwide simultaneous observations of the weather (Chenoweth 1993). Today, weather observations are routinely measured every hour at airfields worldwide and can be sampled much more frequently by automated weather stations. For climatologists, the time of observations is of critical importance. Temperatures recorded daily at the same time are critical to account for changes in the diurnal cycle of temperature from minimum temperatures (usually near sunrise) to maximum temperatures (usually in midafternoon) (Karl et al. 1986). When an observing network does not have a standardized time-of-observation schedule, the climatologist must adjust the individual records so that they conform to a single standard.

Instrument exposure is also of great importance. Thermometers require protection from being wetted by precipitation and from being exposed to direct and scattered solar radiation (Parker 1990). The design of such thermometer screens only became widespread in the United States in the mid-1880s at official government weather stations, and between 1897 and 1903 at the cooperative weather station network sponsored by the government and manned by individual volunteers (Chenoweth 1993). Proper site location for a thermometer screen is in an open area exposed to a free range of air circulation and sunlight, on soils prevalent in the local area, and where possible, in terrain typical of the prevailing area (Sparks 1970). Exposure of rain gauges, wind vanes and anemometers, and other weather instruments are also affected by their site exposure.

Climatologists require metadata (data about the data itself) to sort and classify stations based on their instrument exposure (Hadeen and Davis 1990). A judicious selection procedure can identify the best records for use in long-time series that are least prone to nonclimatic biases and require fewer adjustments to the raw data to make them climatically homogeneous through time and across the geographic area of interest. Also, as much data as can be gathered is necessary in order to reduce the effect of inevitable different exposure qualities between stations and random errors in the data.

Before a climatically homogeneous record can be produced, the following information is ideally available for assessment:

1. The location(s) of the station and the date(s) of any change in location of the weather station. This information should include the place name of the site, the coordinates, site elevation, and elevation of instruments with respect to the ground.

2. The observer(s) at a station and the date(s) of change in the observer(s). At official weather stations, a team of observers with comparable training is assumed, and this information is normally not needed. For stations with one observer, and in all premodern observing networks (generally before the late nineteenth century worldwide), this information is very important because individual observers have unique procedures and habits. If the observer was part of a weather-observing network, this should also be noted.

3. Types of and changes in instrumentation. The date of a change in instrumentation is of critical importance, since instrument performance can vary. If available, the manufacturer and instrument number and calibration information should be recorded.

4. Types of and changes in the exposure and placement of instrumentation at a given location. This information includes the location of instruments with respect to one another and other local features that could affect the performance of the instruments (buildings, walls, dense stands of trees). This includes the type of thermometer screen (if any) used for thermometers, the presence of a windshield around a rain gauge, the distance of the instruments from local features, and the height and dimension of such features.

5. The time of observation(s) and the elements routinely recorded. For climatological purposes, at least one daily reading, made at the same time (usually with a tolerance of plus or minus ten minutes from the nominal observing time), is required. If different weather elements are observed at different times, it is essential that these be specified.

6. Other site-specific information essential for the interpretation of a record. For example, if an observer is frequently absent and has a regular substitute observer, this should be made clear either in a note by the observer or at least be apparent by differences in handwriting.

7. The location of the original manuscript source data, its physical condition, and availability. Also, the location and availability of any published data drawn from this source.

In most cases, not all of this information is available. By working with large numbers of station reports, and using the best-documented station records, there are several statistical approaches that are suited to providing relatively homogeneous climatic records (a perfectly homogeneous record being virtually impossible to create). Basic statistical analysis (means, standard deviations, ratios, frequency distributions, and analysis of trend) are standard tools for the construction of long-term climatic series. The detection of urban heat-island effects, which are often gradual at a given site, is important for temperature reconstruction. The magnitude of urban heat islands can be measured relative to neighboring rural stations, but for detection of natural climatic variability, it is best (when possible) to choose rural locations to minimize the potential for such biases (while still testing for trend). Such a series for the United States currently exists: the U.S. Historical Climatology Network (USHCN), with a record from 1895 to the present (Easterling et al. 1996).

Constructing relatively homogeneous climatic time series requires a trade-off between resources, data availability, and geographical range of interest. The USHCN is a network of over 1,000 individual stations that are constantly updated and recalculated using computer code that runs into tens of thousands of lines. Not all of the available data can be used in such a computationally intensive database. Single-station or small-area networks can make use of all available data, but this requires an intensive effort to locate and digitize the records before any analysis begins.

The construction of a relatively homogeneous climatic time series usually involves the comparison of overlapping time series. Periods of common operation are compared, and the difference is applied to an earlier portion of the record in an iterative procedure back in time until either the beginning of the record is reached, or there are too many actual or potential discontinuities to reliably reconstruct the weather element. Different methods are used based on the available metadata and statistical characteristics of the available data. A good approach is to use all known information on discontinuities to provide an initial "first-guess" series, and then test other stations against this series for suspected or potential discontinuities (Karl and Williams 1987). The USHCN network represents a data set composed from such an approach.

A limitation of most data sets, such as that of the USHCN is that they are limited to mean temperature (derived from the maximum and minimum temperatures) and/or rainfall amounts. Fewer series are available for hourly temperature, barometric pressure, and wind direction and speed.

When working with weather records from the seventeenth through much of the nineteenth century, the problem of data incompatibility increases. In part, this is because much of the historical records are noninstrumental. For example, rainfall totals are rare before the early-to-mid nineteenth century in most regions of the world, but observers frequently made note of the occurrence of precipitation and its relative intensity. The number of days with precipitation is a climatically useful statistic that is not readily available from twentieth century records without the acquisition and processing of considerable amounts of data. Wind and cloud cover data are also common weather elements in historical weather documents. There are challenges in the interpretation of these older records, such as the conversion of wind-force terminology into wind-speed values. However, it is not always safe to assume that modern data are necessarily more accurate or reliable than the older data. For example, wind-speed data are notoriously difficult to homogenize in modern land data because of difficulties in finding comparable exposures at different stations. Rainfall totals, which can be highly variable in space, may not always be as useful as rain days for some studies (Diaz 1991).

The challenge for historical climatologists is to locate and digitize historical daily weather records and identify those available weather elements that provide the maximum amount of information about the climate and the physical atmospheric processes involved in determining the local weather and its longer term variability (Dherent and Petit-Renaud 1994). An informed selection of data, based on modern knowledge of atmospheric and oceanic processes in the Earth's climate system, can assist in choosing historical weather documents for study. Some weather records are from regions sensitive to large-scale atmospheric processes, such as the El Niño/Southern Oscillation (ENSO), the North Atlantic Oscillation, the Asian Monsoon, and other local and regional phenomena. As mentioned in Chapter 1, Jamaica lies under the influence of both the Azores High and the ENSO phenomenon in the tropical Pacific Ocean. Both of these weather regimes are major players in regional and global climate. The Thistlewood record provides the first and best examination of tropical weather at such an early date.

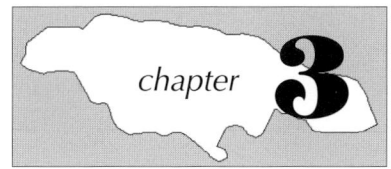

chapter 3

THE THISTLEWOOD WEATHER RECORD

The weather records of Thomas Thistlewood constitute almost one third of the entire manuscript collection of about 15,000 pages, which are preserved at the Lincolnshire Archives, Lincoln, U.K.[4] Nearly 5,000 pages are a daily weather record kept by Thistlewood from 26 July 1750 through 15 November 1786 (all dates are converted to the modern Gregorian Calendar unless otherwise stated) kept near Savanna-la-Mar, Jamaica (18°14′N 78°10′W). The remainder of the collection consists of a journal documenting daily activity at Egypt Plantation and Bread Nut Island Plantation, a commonplace book and miscellaneous papers.

Thistlewood has left the world's earliest continuous daily weather record kept by one person from anywhere outside of Europe and the United States.[5] His rainfall record, kept from 1 July

[4] The Monson Collection, Thistlewood Papers, 31 (Journals and the manuscripts of the Thistlewood family of Lincolnshire and Jamaica) consists of 92 separate pieces. The daily weather entries for 1750–1751 are found in 31/2 (before the weather journal was separated from the daily activity journal), and 31/39 through 31/72 for the years 1752–1786. Hereafter, I refer to the journals as "Thistlewood, Monson Collection 31/piece number." The journals are in daily chronological sequence so the references can easily be found in the original by date.

[5] The earliest known English language weather diary from Jamaica was kept by Sir Hans Sloane in 1688 at Spanish Town (Long, 1774, pp. 646–648). In India, G.E. Geisler kept a weather record, at Madras, from October 1732 to May 1737 (Glaser et al. 1991). Sloane and Geisler had no instruments except possibly wind vanes. Early travel accounts probably account for the earliest temperature observations from Carthagena (1735), Panama and Ecuador (1736), Curaçao (1760), St. Eustatia (1760), and Sierra Leone (1763) noted in Long (1774). Long also makes reference to rainfall observations in Surinam and Barbados that date probably from the mid-eighteenth century. A four-year record of rainfall was kept at Antigua from 1751–1754 (Byam 1755). The earliest known temperature readings made at sea date from July 1760, in latitude 18ºN (Long, 1774, p. 657). The earliest long-term record in the United States, kept at Charleston, South Carolina from 1738–1750 is only available in abstract form. The original daily record is apparently lost (Cary Mock, personal communication).

1760 through 15 November 1786, is the world's oldest instrumental rainfall record of this length from the Tropics. Some important details about his observing practices are answered in part from correspondence and data provided to Sir Edward Long[6], author of *The History of Jamaica* (Long 1774).

Thistlewood was born on 21 March 1721 in Lincolnshire, England. The surviving papers of Thistlewood begin with his return from India as a sailor in the employ of the English East India Company. This voyage took place from 1746 to 1748. Until May 1750, Thistlewood spent most of his time in England; during this time, now almost thirty, he decided to make his life in the New World, and sailed to Jamaica, arriving in Savanna-la-Mar in May 1750. In July 1750, he began a year's employment near Black River, Jamaica, acting as overseer for a slave plantation. He quit this job in mid-July 1751 and returned to Savanna-la-Mar. On 27 September 1751 he accepted the offer of work as an overseer in the employ of James Dorrill at Egypt Plantation, where he lived and worked until 1767. In 1765, Thistlewood purchased the former Paradise Pen, located about one mile east and northeast of Egypt Plantation (Hall 1989). He did not occupy his new home until 4 September 1767, and renamed it Bread Nut Island. He died at his home on 30 November 1786. Figure 1 depicts the locations of his residences in Jamaica.

Thistlewood kept a combined journal of daily activity and weather in 1750 and 1751. After leaving Black River in July 1751, there are only a few mentions of the weather until 1 January 1752 (Old Style). By this time, he had been working for three months as overseer at Egypt Plantation, northwest of the town of Savanna-la-Mar. From 12 January 1752 (1 January Old Style) he separated his weather record from the daily activity journal, and kept this practice for the remainder of his life.

Appendix 1 provides all metadata for the three sites at which Thistlewood kept his record in accordance with the information on station metadata described in Chapter 2. The weather record, as first kept in 1750, included observations of the presence of fog or haze, state of sky, occurrence of rain and thunderstorms, and the prevailing wind force. Wind direction was mentioned only with reference to tropical storms and hurricanes and for winter month "North's," which are incursions of cooler and much drier air from North America that are a major part of the winter dry season weather regime in Jamaica. Also included in his weather record are

---

[6] British Library, Manuscripts Section, Edward Long Collection, Add. Ms. 18275A, ff. 128–128v). Hereafter referred to as Long Collection.

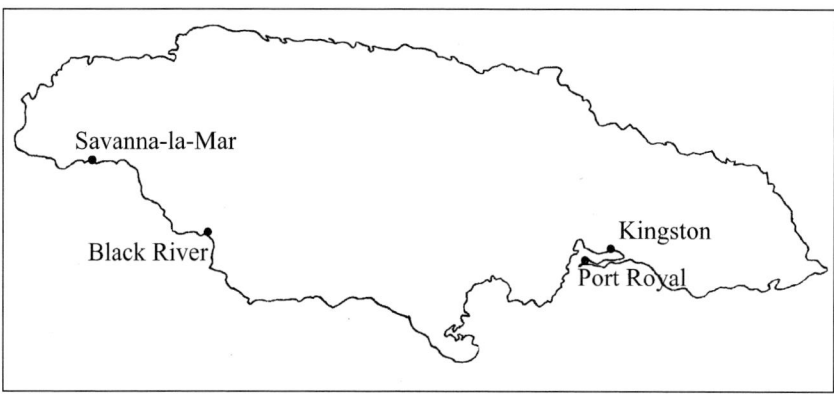

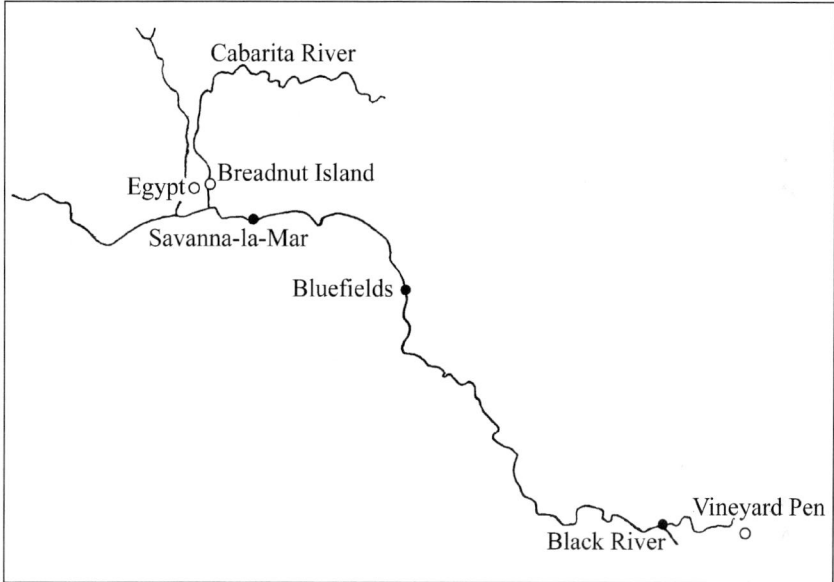

FIG. 1 Maps depicting the locations where Thistlewood resided from 1750 to 1786.

brief descriptions of lunar and solar eclipses, meteors, and earth-quakes. In May 1753, Thistlewood began to include temperature readings from an apparent Hauksbee-type thermometer (the thermometer scale was inverted from modern thermometers, with warmer temperatures at lower values of the scale) (Middleton 1966). This record continued until June 1754 but was stopped, apparently due to the poor performance of the instrument. Beginning in January 1760, Thistlewood began to include wind direction in his daily records and continued this practice for the remainder of the record. In July 1760, daily rainfall totals appear and continue for the remainder of his weather record. Thistlewood may at the same time have purchased a Fahrenheit's thermometer, for his first mention of a temperature reading using the Fahrenheit scale occurs in 1761. From 1764 to 1767, his record includes separate sheets containing temperature readings made daily at sunrise, noon, and sunset. He continued making his readings after 1767, but daily records for 1 January 1770 to 7 June 1776 are the only remaining extant data.[7] As shown below, the change in observing practices in 1760 mark a major discontinuity in observing practices by Thistlewood to more precise and accurate observations.

ANALYSIS OF THISTLEWOOD'S OBSERVING PRACTICES

All historical weather records must be evaluated for their potential usefulness in reconstruction of past weather and climate. Premodern weather observations were made in a variety of locations with unique exposures, and were written in personal diaries, ships' logbooks, mission accounts, published in newspapers, or written in the margins of almanacs, making for a variegated lot of data that are of little use unless critically studied by climatologists. Where discrete numerical information is available, careful screening of a record with simple graphical methods such as scatter diagrams and frequency distribution tables serve as powerful analytic tools. Modern climatological data, when available, also help in the interpretation of historical records by setting realistic limits on the statistical properties of a given weather element. Data analysis methods such as that advocated by Tukey (1977) provide a useful framework for analysis. It also provides a more complete description of the local climate as recommended by Guttman (1989).

As a private observer, Thistlewood's observing methods and procedures are unique, but an examination of his record reveals

---

[7] Long Collection, Add. Ms. 18275B, ff. 64.

important influences. I constructed a digital database of the entire weather record,[8] designed to best fit his unique practices. The most effective way to speedily and to accurately input his data was to take each time increment and include all the data (which varies from observation to observation) on one record line. Each line contains information that contains information identified in the columnar headings. The columnar headings are listed in the order they appear in the database.

> **YYYY** is the year.
> **MMDD** is the month and day (read March 1 as 301; December 15 as 1215, etc.).
> **Hd** is the observation time modifier (see Appendix 2).
> **HH** is the "Time of Day" (TOD) or observation time period.
> **DDD1** is the prevailing wind direction; if only one wind direction is given, no value is entered in the succeeding column.
> **DDD2** is the second prevailing wind direction when stated; if more than two directions are given, only the first and last in the sequence are used.
> **FF** is the wind force term.
> **N** is the state of sky.
> **WW1** is the prevailing present significant weather, if any. This field is blank when there is no significant weather stated (see Appendix 3).
> **WW2** is the second prevailing present significant weather, if any. This field is blank when there is only one or no significant weather (see Appendix 3).
> **FRR** is the Rainfall Frequency descriptor.
> **IRR** is the Rainfall Intensity descriptor.
> **TRR** is the Rainfall Type descriptor.
> **RRR** is the amount of rainfall in inches and hundredths of inches.
> **FROM** is the beginning time of a precipitation or significant weather event, when stated.
> **TO** is the ending time of a precipitation or significant weather event, when stated.

---

[8] The daily temperature data made at sunrise, noon, and sunset are not included. This is because the design of the database was made when only the 1753–1754 and 1764–1767 data were known to exist. Also, the daily temperature readings are separate from the main weather journal. The poor quality of the 1753–1754 data excluded it from consideration for digitization. Subsequent discovery of the 1770–1776 data came after the initial design. For this reason, these data are included in a separate spreadsheet file. Sporadic temperature data otherwise mentioned are included in the database.

> **GR** is the "grouped rainfall" descriptor/rainfall amount categorization value used to reduce the eighty-two unique rainfall descriptors into eight categories that best correlate with actual rainfall amounts.
>
> **TEMP** is a subjective estimate of temperature and humidity and some isolated instrumental temperature values.
>
> **TEXT** is the gist or direct quotations from the weather journal that are not categorized or included for clarity of the encoded values.

Thistlewood used nineteen different time-of-day (TOD) descriptors for his observations. Table 1 gives the monthly frequency of TOD descriptors. The most common are "A.M." (usually unstated but assumed based on its appearance in days when an earlier observation preceded and was followed by an "A.M." observation) and "P.M." he describes them as:

> A.M. stands for the Ante Meridian. Forenoon.
> P.M. for Post Meridian. Afternoon.[9]

As can be seen from Table 1, he used the TOD "afternoon" only 51 times, but "P.M." 7,917 times. By his definition, these observation times can be considered equivalent. Other TODs are less specific, although the meaning is clear, and I have listed the TODs in what appears to be the time sequence that is meant by Thistlewood, from the context of the written weather descriptions. At night, when Thistlewood would normally be asleep, there are fewer observations, as expected. A seasonal distribution of his "Day" TOD (sunrise to sunset) shows far more such TODs in the winter dry season than during the summer rainy season. This reflects the normal climate of this location. In the winter months, dry sunny weather prevails throughout the day and the sea breeze often does not appear. During the summer, the day begins sunny, but by afternoon the sky is mostly cloudy and afternoon thunderstorms form and drop rain. The sea breeze is an almost daily event. The increase in the number of "A.M." and "P.M." observations from the winter to the summer months is opposite to that of the "Day" TODs. The "At Night" TOD is unique to Thistlewood. This observation time, which seems to be made just before he retired each evening, is virtually always used for describing distant lightning and/or thunder. The increase in frequency of this term from winter to summer is another reflection of the local climate. Thunder and lightning are much more frequent in the rainy season than during the dry season. Thistlewood describes the typical diurnal cycle of rainfall:

[9] Thistlewood, Monson Collection, 31/89, p. 58.

TABLE 1.

Monthly frequency of time of Day (TOD) descriptors used by Thistlewood in his weather journal. TODs are listed in time sequence, except for "Day" (bottom), which refers to the period from sunrise to sunset.

TIME PERIOD	MONTH												
	Jan	Feb	Mar	Apr	May	Jun	Jul	Aug	Sep	Oct	Nov	Dec	YEAR
Midnight	0	1	0	1	0	1	1	1	1	0	2	0	8
Before day	1	4	3	0	0	0	0	3	7	9	10	3	40
Morning	133	102	162	142	130	125	81	56	56	69	78	60	1194
Sunrise	0	0	0	5	2	4	5	8	0	8	3	2	37
A.M.	501	535	593	619	774	810	929	924	860	826	565	489	8425
Forenoon	1	1	0	0	0	0	0	1	0	2	0	2	7
Forepart	1	0	0	0	0	0	0	0	0	1	0	0	2
Towards noon	9	15	19	22	20	14	34	30	23	25	7	17	235
Noon	26	36	24	24	41	25	32	23	28	35	14	19	327
Middle of day	56	61	73	86	79	125	79	60	51	49	28	13	760
P.M.	472	477	555	570	729	765	873	885	820	760	536	475	7917
Afternoon	4	2	5	3	8	5	5	3	4	7	3	2	51
Towards eve.	0	1	1	2	0	1	2	1	1	1	0	0	10
Evening	317	266	296	288	309	307	267	255	264	276	269	230	3344
Sunset	2	1	2	3	0	3	4	2	5	10	2	6	40
At night	97	74	81	104	383	499	496	525	529	516	235	143	3682
In the night	16	10	7	13	10	5	7	6	8	12	19	16	129
All night	1	1	4	6	8	6	3	3	6	7	5	2	52
Day	504	404	400	354	239	165	111	128	159	231	436	540	3671

In this part it is not very common to have rain in the Forenoon, our heavy showers usually coming on at 1, 2, or 3 o'Clock in the afternoon, sooner or later as it happens; though when the wind deviates from the Trade, that is gets too much to the South, or North (for I have hardly ever known it from the West), we have then rain indifferently at any time in the day or night . . . .[10]

With Thistlewood's normal observing times at hand, it is possible to determine those observations most frequently used. For the entire record kept near Savanna-la-Mar (1752–1786), there are 12,772 daily records. The total number of observations exceeds the number of days because there are frequently multiple TODs per day. If only one observation per day is given, the TOD is invariably "Day." If two observations per day are recorded, it is most frequently "A.M." and "P.M." or "Day" and "At Night." The combinations increase with the number of TODs.

This information is useful for limiting the database manipulation and extraction of data. Because most of Thistlewood's observing times are for a block of time, and not usually a specific moment, this makes his record (and most such historical records) different from modern observing weather networks, which normally provide hourly data made at specified times. His record should be thought of as the "prevailing" or "average" weather conditions during each TOD, with certain exceptions when he is more specific.

INDIVIDUAL WEATHER DESCRIPTORS

Cloud Cover and Sunshine

Thistlewood used eight descriptors to describe the state of the sky; they are listed in Table 2. While the terms "clear" and "cloudy" would seem apparent, the rare usage of "clear" and the more ambiguous "fair and clear" in favor of "fair" is revealing. Again, with reference to the modern climatologies of cloud cover, there is a difference between the everyday usage of "fair" by Thistlewood and the observing procedures for determining cloud amount and types in modern data.

Four distinct cloud genera are internationally recognized: (1) High-level clouds (on average more than 15,000 feet above ground level); (2) Middle-level clouds (between 8,000 and 15,000 feet); (3) Low-level clouds (cloud ceilings below 8,000 feet); and (4) Clouds with vertical development (including most clouds producing thunder and lightning) (Berry et al. 1945). Multiple cloud

---

[10] Thistlewood, Monson Collection, 31/89, p. 62.

TABLE 2

Monthly frequency of sky cover terms in the Thistlewood weather record, 1752–1786. (A) All terms (B) Combined terms using the first five descriptors in A for "Fair" and the remaining three descriptors for "Cloudy". The higher frequency of cloudy weather in May–October reflects the normally cloudier weather of the rainy season in Jamaica, while the higher frequency of fair weather from November through April reflects the sunnier weather of the dry season.*

A

SKY DESCRIPTOR						Month						
	Jan	Feb	Mar	Apr	May	Jun	Jul	Aug	Sep	Oct	Nov	Dec	Year
Clear	2	0	1	0	1	0	0	0	2	1	1	2	10
Fair and Clear	0	3	4	1	0	3	0	0	0	0	1	1	13
Fair	232	194	228	180	160	140	165	93	74	90	127	206	1889
Most part fair	93	67	105	112	62	63	48	49	52	52	90	103	896
At times cloudy	720	696	756	718	744	727	780	811	780	756	718	733	8939
Most part cloudy	158	141	172	146	161	128	154	177	151	185	155	128	1856
Somewhat cloudy	31	19	14	27	22	44	35	33	25	23	15	14	302
Cloudy	660	700	718	765	1053	979	1028	1042	1097	1097	724	573	10436

B

SKY DESCRIPTOR						Month						
	Jan	Feb	Mar	Apr	May	Jun	Jul	Aug	Sep	Oct	Nov	Dec	Year
Fair	1047	960	1094	1011	967	933	993	953	908	899	937	1045	11747
Cloudy	849	860	904	938	1236	1151	1217	1252	1273	1305	894	715	12575

*Shortly before the publication of this book, evidence was found that allowed the inclusion of a tropical storm on 28 September 1791 to Table 23. However, Figures 10 and 11 do not include this data point.

types are frequently present, and a sky is considered overcast if the entire sky is covered with clouds, despite their appearance, thickness, or ability to diminish incoming sunlight.

The first attempts at classifying clouds for scientific studies were not published until the beginning of the nineteenth century. Therefore, Thistlewood's descriptive terms should not be considered to be a faithful record of cloud amount and type as understood today. He defines some of his terms:

> By fair weather is meant when the sky is serene & clear, few or no clouds in sight. By 'at times cloudy' is meant when but few small clouds, the Sun not being clouded perhaps half an hour in the whole day; this is our most usual weather, though in this sort of weather we also frequently see great congregated bodies of heavy clouds at a distance over the Mountains, where they empty themselves in plentiful showers . . . . [11]

With these self-described definitions it is not a great leap to infer a definition for "most part cloudy" that is an approximate mirror image of "at times cloudy". For cloudy skies, I have included all uses of this word and also all instances in which rain or drizzle was falling, since Thistlewood never uses any sky cover term when precipitation is falling. I assume that the overwhelming numbers occur with a cloudy sky, although there are instances when the sun can be shining during a rain shower. In the database, "c" stands for cloudy as stated by Thistlewood, while "x" stands for cloudy as inferred from the presence of precipitation. The term "somewhat cloudy" is perhaps the most ambiguous. The term is more frequent in the rainy season, and I interpret it to indicate instances of multiple cloud layers in which the sun is generally diminished by a high-level cirrostratus cloud deck, but lower-altitude clouds are also present but not constantly obscuring the sun. The term "hazy sunshine" may be apt.

Cloud terms were placed into two groups. The first five rows of Table 2 were merged together to form a single "Fair" category and the remaining three rows into a single "Cloudy" category. More precisely, the two categories should be considered as (1) completely or mainly sunny (with or without accompanying clouds), and (2) little or no blue sky and, therefore, little or no sunshine. This would fit a more common distinction that would be made by the average person even today. Is the sun shining or not?

Table 2B lists the monthly frequency of the merged terms. Fair weather is more frequent than cloudy weather during the dry season and cloudy weather is more frequent than fair weather in the

---

[11] Thistlewood, Monson Collection, 31/89, pp. 64–65.

rainy season. I used a simple binary encoding scheme (1 equals sunny; 0 equals not sunny) to produce monthly mean sunshine indices ranging from 1.00 for a completely sunny month to 0.00 for a completely cloudy month.

The Thistlewood data are sensitive to sunshine amount, whereas the modern record of cloud cover is not directly comparable with sunshine amount. The relationship between cloud amount and sunshine is not a direct one in the modern data, because some cloud types diminish the sunlight more than other cloud types. Historical weather records of cloudiness and sunshine are distinct from the modern records; only by comparing independent records that operated simultaneously can we attempt to create long records of cloudiness and sunshine variability. However, we can make relative assessments for individual records such as Thistlewood's to determine year-to-year variability and also as a check on other weather elements in his record.

On this last point, Table 2 shows more instances of cloudy weather in the rainy season than in the dry season. This fits with the climatology of the island. The frequency of multiple TODs was higher in the rainy season. This is seen in the frequency of the "Day" TOD at the bottom of Table 1. This is also a reflection of the greater diurnal variability of the weather in the rainy season. Both Thistlewood's choice of time intervals (reflecting his perception of significant differences in the prevailing weather in each time interval) and his actual observations of sunshine and cloudiness are internally consistent and lead to greater confidence in the quality of his observations.

Rainfall

Thistlewood made note of every instance of rain and drizzle he observed, ranging from a few sprinkles of rainfall to torrents of hurricane-driven rain. Table 3 provides the encoded values for each of the three parts comprising the rainfall frequency/intensity/type descriptors. Table 4 lists the eighty-two rainfall descriptors used and their frequency of use. Twenty-six descriptors were used only once. The database also includes a separate line record labeled "SUM" to designate the 24-hour precipitation total.

In Jamaica, rain and rain showers, with or without thunder and lightning, are most common. Drizzle is also observed. All days with only one rain event that could be confidently equated to the rainfall recorded in the rain gauge were matched against each of the eighty-two descriptors. This was used to sort the terms into eight major categories of rain events, ranging from the lightest to the

Text continues on p. 28

TABLE 3

Rainfall frequency, intensity and type descriptors used in the Thistlewood weather record and their equivalent data base encoded values. An example, "Some good showers of rain" is encoded as 5,3,4, with the '5' from the FRR, the '3' from the IRR, and the '4' from the TRR.

Rainfall Frequency Descriptor (FRR)	FRR Encoded Integer Value	Rainfall Intensity Descriptor (IRR)	IRR Encoded Integer Value	Rainfall Type Descriptor (TRR)	TRR Encoded Integer Value
No modifier	0	No modifier	0	Drizzle	0
Not used	1	Drops of	1	Drizzling rain	1
A, an	2	Small	2	Showers of drizzling rain	2
Some few	3	Good	3	Not used	3
At times	4	Pretty good	4	Showers of rain	4
Some	5	Moderate	5	Rain	5
Several	6	Hard	6	Rain with squalls of wind	6
Frequent	7	Very hard	7		
Most part	8	Prodigious (exceeding hard)	8		
Almost continual	9				
Continual	10				

TABLE 4

Rainfall Event Descriptors, their equivalent encoded data base values, and frequency of occurrence. The final column groups the individual descriptors into eight classes that are correlated with the total amount of rainfall. The higher the grouped rainfall value, the greater the total rainfall.

Rainfall Event Descriptor	Rainfall Frequency FRR	Rainfall Intensity IRR	Rainfall Type TRR	Frequency of Event Descriptor	Grouped Rainfall (see Table 5) GR
A few drops of rain	2	1	5	1007	1
Some few drops of rain	3	1	5	123	1
At times drops of rain	4	1	5	16	1
Some drops of rain	5	1	5	2	1
A small drizzle	2	2	0	7	1
A drizzling rain	2	2	1	1	1
A small shower of drizzling rain	2	2	2	2	1
Drizzle	0	0	1	33	2
Drizzling rain	2	0	1	3	2
At times drizzling rain	4	0	1	29	2
Some drizzling rain	5	0	1	182	2
Frequent drizzling rain	7	0	1	2	2
Some showers of drizzling rain	5	0	2	1	2
Some small showers of drizzling rain	5	2	2	2	2
At times small showers of drizzling rain	4	2	2	1	2
A small shower of rain	2	2	4	901	3
Frequent showers of drizzling rain	7	0	2	4	4
Small showers of rain	0	2	4	2	4
Some few small showers of rain	3	2	4	1	4
At times small showers of rain	4	2	4	3	4
Small rain	0	2	5	3	4
A small rain	2	2	5	1	4
Small rain (with squalls of wind)	2	2	6	3	4
Rain (with squalls of wind)	0	0	6	4	4
Some small rain (with squalls of wind)	5	2	6	1	4
At times rain	4	0	5	1	4
Some rain	5	0	5	28	4

(continued)

Table 4 *(continued)*

Rainfall Event Descriptor	Rainfall Frequency FRR	Rainfall Intensity IRR	Rainfall Type TRR	Frequency of Event Descriptor	Grouped Rainfall (see Table 5) GR
Some moderate rain	5	5	5	29	4
Most part drizzling rain	8	0	1	7	4
Almost continual drizzling rain	9	0	1	1	4
Continuous drizzling rain	10	0	1	3	4
Several small showers of rain	6	2	4	9	4
Frequent small showers of rain	7	2	4	9	4
A shower of rain	2	0	4	20	5
At times a shower of rain	4	0	4	2	5
Some showers of rain	5	0	4	208	5
Some moderate showers of rain	5	5	4	4	5
A moderate shower of rain	2	5	4	18	5
Moderate showers of rain	0	5	4	1	5
A pretty good shower of rain	2	4	4	3	5
Moderate rain	0	5	5	61	5
A squall of wind with rain	2	0	6	61	5
Some rain and squalls of wind	5	0	6	21	5
At times rain and squalls of wind	4	0	6	1	5
Several showers of rain	6	0	4	38	6
Frequent showers of rain	7	0	4	36	6
Most part moderate rain	8	5	5	17	6
Very hard rain	0	7	5	3	6
Rain	0	0	5	36	6
Several squalls of wind with rain	6	0	6	1	6
Frequent squalls of wind with rain	7	0	6	8	6
A good shower of rain	2	3	4	70	6
Some good showers of rain	5	3	4	8	6
Several good showers of rain	6	3	4	1	6
Continuous moderate rain	10	5	5	4	6
Almost continual moderate rain	9	5	5	6	6
Frequent moderate rain	7	5	5	1	6
Some hard rain (with squalls of wind)	5	6	6	6	6
A good rain	2	3	5	1	6
A good deal of rain	2	4	5	3	6
A very fine moderate rain	2	5	5	1	6
A hard shower of rain	2	6	4	193	7
A hard rain	2	6	5	1	7
A very hard rain (with squalls of wind)	2	7	6	1	7
Some hard showers of rain	5	6	4	28	7
Several hard squalls of rain	6	6	6	1	7
Most part rain	8	0	5	10	7
Almost continual rain	9	0	5	20	7
Continuous rain	10	0	5	23	7
Hard showers of rain	0	6	4	1	7
(A) very hard shower of rain	0	7	6	2	7
A prodigious (or exceeding hard) shower of rain	2	8	4	3	8
As above (with squalls of wind)	2	8	6	1	8
A very hard shower of rain	2	7	4	5	8
Continual showers of rain	10	0	4	1	8
Continual hard rain	10	6	5	1	8
Nearly continuous hard rain	9	6	5	1	8
Several pretty good showers of rain	6	4	4	1	8
Several very hard showers of rain	6	7	4	1	8
Several prodigious (or exceeding hard) showers of rain	6	8	4	1	8
Several hard showers of rain	6	6	4	2	8
Most part hard rain	8	6	5	3	8

heaviest falls of rain. Table 5 lists the rainfall groupings and statistical features of each group.

The groups in Table 5 represent precipitation events that approximately double in average amount per category. Group 9 is taken only from hurricanes and tropical storms affecting western Jamaica. Despite the small sample size, the amount is roughly double that of group 8.

Each of the eight groups is summed to produce monthly and annual totals for the entire period of record. Table 6 presents the annual total for each group for the period of record. Groups 1, 2, 3, 5, 7, and 8 show distinct and sudden changes in their frequency after the 1750s. Table 7 shows each group's contribution to the total number of annual rain days, expressed as a percentage of the total number of rain days included in all eight groups. The most striking discontinuity is at the beginning in July 1760 with group 1. There is an unprecedented move to zero occurrences in exactly the month that Thistlewood began using a rain gauge to measure rain amount. The relative contribution of each category (Table 6) indicates that rainfall groups 1, 7, and 8 were used by Thistlewood much more often before 1760 than afterwards, whereas groups 2, 3, and 5 were used much less often before 1760.

Table 7 includes the cumulative frequency of each category for the period 1752–1759 and 1761–1786 (1760 is excluded due to the apparent discontinuity that year). With reference to Table 5 and the average rainfall amounts for each of the eight groups, it is apparent that Thistlewood became more precise in his definition of "drops of rain" (Group 1) after the 1750s. Many light measurable rains were described as "drops of rain" or drizzle in the 1750s. The frequency of usage is about seven times higher in the 1750s than afterwards. Similarly, in group 8, the most intense and persistent rains were more than four times as frequent in the 1750s than subsequently.

With reference to Table 5, where each category's average rainfall amount was given, the application of these values to the 1750s led to very high, and likely overestimates, of monthly and annual rainfall. As will be discussed later, such high rainfall totals are inconsistent with other features of the record and other descriptions of the weather during documented severe droughts.

The discontinuity in the frequency of certain descriptors of rainfall intensity and frequency is best explained by a change in procedures and perceptions by the observer. Beginning in January 1760, Thistlewood recorded the wind direction along with the wind force on a daily basis. In July 1760, daily rainfall measurements began. In a letter to Edward Long, Thistlewood wrote, ". . . . have at different times fitted up [a] pluviometer for some of my acquaintance to observe the quantity of rain that fell in different parts of the parish, which they have promised carefully to notice, but could never get an account for a week together that could be depended on".[12] This comment is the only insight into the rain gauge that he used. Thistlewood himself apparently manufactured it.

The greater care and quantification of weather elements indicates a deepening interest in accurate weather observations. Although Thistlewood does not explain his reasons, his horticultural interests, which involved the growing and selling of local and exotic plants to his neighbors (Hall 1989) were probably contributing factors. His June 1777 letter to Edward Long discusses various plantings and the effect of the weather on their cultivation. Furthermore, several years of close observation of the local weather led him to scrutinize, and more accurately categorize, his noninstrumental estimates of rainfall intensity. It is noteworthy that the shift in category frequencies is very nearly identical with the beginning of the instrumental readings, and did not (except for category 1, where there is a second smaller discontinuity in 1764) lag the instrumental readings. Had they lagged, then his estimates of rainfall intensity may have been modified by his rain gauge readings. Instead, the near synchronicity of the changes argues for his perceptions and procedures to have changed at the same time. In short, a good observer became an excellent observer from 1760 onward.

The changes in the rainfall categories show the importance of content analysis techniques to historical weather records. The

TABLE 5

Statistical summary of the main rainfall groupings (see Tables 3 and 4). N is the number of observations. X is the average rainfall amount where 'T' indicates a trace of rain, or less than .005 inch of rain. The 10th, 50th (median) and 90th percentile values are given in the final three columns. The rainfall amounts roughly double in amount from one group to the next. Tropical cyclones produce about twice as much as other heavy rains in Group 8.

	N	X	10%	50%	90%
Group 1	1159	T	T	T	0.01
Group 2	274	0.03	T	0.03	0.07
Group 3	893	0.06	0.02	0.07	0.12
Group 4	297	0.14	0.04	0.12	0.29
Group 5	1191	0.36	0.08	0.29	0.81
Group 6	231	0.7	0.26	0.61	1.17
Group 7	323	1.26	0.63	1.18	2.1
Group 8	20	2.61	1.57	2.35	3.74
Tropical Cyclones	18	4.36	1.86	4.48	8.27

[12] Long Collection, Add. Ms. 18275A, ff. 122–124.

TABLE 6

Annual number of days with rainfall by grouped rain categories (see Table 4) for each year from 1752 to 1786. These figures are converted into percentage frequency in Table 7 to highlight the changes over time in their relative frequency with one another, indicating a change in observing practices by Thistlewood. These changes introduce a non-climatic trend into the data that must be accounted for before the data can be useful for climatological purposes.

| | | | | RAINFALL GROUP | | | | | |
Year	1	2	3	4	5	6	7	8	SUM
1752	57	6	28	21	43	36	31	1	223
1753	44	7	30	24	54	16	39	1	215
1754	49	9	17	24	32	18	25	0	174
1755	68	4	28	16	27	28	24	8	203
1756	48	4	26	16	35	18	22	5	174
1757	45	11	31	21	63	24	53	8	256
1758	62	6	23	19	37	21	24	3	195
1759	50	5	36	20	42	30	18	3	204
1760	44	6	36	10	33	21	12	1	163
1761	28	18	33	15	48	18	14	2	176
1762	45	7	33	15	31	9	13	1	154
1763	52	11	34	13	63	11	22	0	206
1764	46	16	34	4	38	15	16	1	170
1765	48	9	31	8	38	14	10	0	158
1766	54	6	36	9	38	17	20	2	182
1767	43	5	32	20	71	15	12	1	199
1768	33	8	40	18	48	12	8	0	167
1769	40	9	31	8	64	10	8	0	170
1770	51	10	27	15	44	14	6	0	167
1771	47	12	25	16	47	19	6	0	172
1772	36	10	38	24	50	19	10	0	187
1773	53	9	29	12	55	8	13	0	179
1774	57	15	24	19	49	23	16	0	203
1775	48	11	40	16	70	19	20	0	224
1776	52	14	59	13	39	12	17	1	207
1777	50	7	38	7	55	17	17	0	191
1778	41	11	49	12	40	25	14	1	193
1779	51	11	38	14	46	23	13	0	196
1780	40	7	36	10	47	27	9	3	179
1781	47	4	44	14	52	28	12	1	202
1782	52	8	32	7	46	23	11	1	180
1783	38	1	33	10	53	26	13	0	174
1784	35	5	37	10	49	28	10	0	174
1785	54	13	34	14	40	28	6	0	189
1786	39	2	21	10	38	25	6	1	142

TABLE 7

Annual percentage frequency of occurrence of rainfall groupings (see Table 4) for each year from 1752 to 1786. A change in observing procedures in 1760 produced a discontinuity in the record that is highlighted in bold and underlined values. The cumulative percentages shows that lighter rainfall amounts were over-represented in the 1750s when compared with the subsequent years.

PERCENTAGE OF ANNUAL OBSERVATIONS FOR EACH CATEGORY									
Year	1	2	3	4	5	6	7	8	SUM
1752	0.26	0.03	0.13	0.09	0.19	0.16	0.14	0.00	1
1753	0.20	0.03	0.14	0.11	0.25	0.07	0.18	0.00	1
1754	0.28	0.05	0.10	0.14	0.18	0.10	0.14	0.00	1
1755	0.33	0.02	0.14	0.08	0.13	0.14	0.12	0.04	1
1756	0.28	0.02	0.15	0.09	0.20	0.10	0.13	0.03	1
1757	0.18	0.04	0.12	0.08	0.25	0.09	0.21	0.03	1
1758	0.32	0.03	0.12	0.10	0.19	0.11	0.12	0.02	1
1759	0.25	0.02	0.18	0.10	0.21	0.15	0.09	0.01	1
1760	0.27	0.04	0.22	0.06	0.20	0.13	0.07	0.01	1
1761	0.16	0.10	0.19	0.09	0.27	0.10	0.08	0.01	1
1762	0.29	0.05	0.21	0.10	0.20	0.06	0.08	0.01	1
1763	0.25	0.05	0.17	0.06	0.31	0.05	0.11	0.00	1
1764	0.27	0.09	0.20	0.02	0.22	0.09	0.09	0.01	1
1765	0.30	0.06	0.20	0.05	0.24	0.09	0.06	0.00	1
1766	0.30	0.03	0.20	0.05	0.21	0.09	0.11	0.01	1
1767	0.22	0.03	0.16	0.10	0.36	0.08	0.06	0.01	1
1768	0.20	0.05	0.24	0.11	0.29	0.07	0.05	0.00	1
1769	0.24	0.05	0.18	0.05	0.38	0.06	0.05	0.00	1
1770	0.31	0.06	0.16	0.09	0.26	0.08	0.04	0.00	1
1771	0.27	0.07	0.15	0.09	0.27	0.11	0.03	0.00	1
1772	0.19	0.05	0.20	0.13	0.27	0.10	0.05	0.00	1
1773	0.30	0.05	0.16	0.07	0.31	0.04	0.07	0.00	1
1774	0.28	0.07	0.12	0.09	0.24	0.11	0.08	0.00	1
1775	0.21	0.05	0.18	0.07	0.31	0.08	0.09	0.00	1
1776	0.25	0.07	0.29	0.06	0.19	0.06	0.08	0.00	1
1777	0.26	0.04	0.20	0.04	0.29	0.09	0.09	0.00	1
1778	0.21	0.06	0.25	0.06	0.21	0.13	0.07	0.01	1
1779	0.26	0.06	0.19	0.07	0.23	0.12	0.07	0.00	1
1780	0.22	0.04	0.20	0.06	0.26	0.15	0.05	0.02	1
1781	0.23	0.02	0.22	0.07	0.26	0.14	0.06	0.00	1
1782	0.29	0.04	0.18	0.04	0.26	0.13	0.06	0.01	1
1783	0.22	0.01	0.19	0.06	0.30	0.15	0.07	0.00	1
1784	0.20	0.03	0.21	0.06	0.28	0.16	0.06	0.00	1
1785	0.29	0.07	0.18	0.07	0.21	0.15	0.03	0.00	1
1786	0.27	0.01	0.15	0.07	0.27	0.18	0.04	0.01	1
1752–59	0.26	0.03	_0.13_	0.10	_0.20_	0.12	**0.14**	**0.02**	
1761–86	0.25	0.05	_0.19_	0.07	_0.26_	0.10	**0.07**	**0.00**	
CUMULATIVE									
1752–59	0.26	0.29	0.43	0.53	0.73	0.85	0.98	1.00	
1761–86	0.25	0.30	0.49	0.56	0.83	0.93	1.00	1.00	

systematic change made by Thistlewood is only apparent by quantifying the frequency of his usage of certain words and phrases. If even as good an observer as Thistlewood can make such major changes in a record, this should stand as a precaution for the interpretation of other records. Many observers are less systematic and precise than Thistlewood. Without knowledge of the observer and the context in which the record is kept, along with content analysis and basic statistical methods to examine the record, an accurate history of the weather cannot be constructed.

Wind Force and Wind Direction

Thistlewood used thirteen different terms for the relative wind force. I have merged into single categories "light airs" with "light airs at times calm" (and minor variations on the category modifier) into one "light airs" category. The terms "light wind" and "light breeze" are taken as synonymous. This is supported by usage given in the *Oxford English Dictionary*. Similarly, "moderate breezes" and "moderate gales" are merged together and the twenty-one records of "strong breeze" are merged with the 514 records of "fresh gales." During the winter months, Thistlewood identified outbreaks of cooler and drier air from the North as "moderate North," "strong North," and "very strong North." Most such instances predate 1760 (when he did not normally record the actual wind direction). I interpret these three terms, respectively, as "moderate breezes," "fresh breezes," and "fresh gales," based on their relative frequency of occurrence. Thistlewood writes that most often a North was felt as a fresh breeze. After combining equivalent terms there remained eight distinct wind force terms (Table 8).

The modern climatology of Jamaica shows generally higher wind speeds in the winter dry season than during the summer rainy season. Table 8 shows "moderate breezes" to be most frequent in the winter months while lighter "light breezes" are more frequent during the summer months. Also, light airs show a greater frequency during the summer months.

Thistlewood's terminology for wind force is at variance with usage seen in other historical records from the eighteenth century. Several different wind force scales were in use at this time. Lamb and Frydendahl (1991) listed common terms in use by English sailors around 1700 that only match with Thistlewood's terminology in three instances (Table 9). The well-known Beaufort scale was not created until 1805 and was a codification of the then-existing usage in the U.K. Royal Navy with some modification (Chenoweth 1999).

TABLE 8

Monthly frequency of wind force descriptors in the Thistlewood weather record, 1752–1786. The counts are based on only the mid-day observations used to generate monthly and annual averages. Additional wind force descriptors, up to hurricane strength, are included in the weather record. This table provides the wind force terms that form the overwhelming majority of terms used by Thistlewood.

WIND FORCE	Jan	Feb	Mar	Apr	May	Jun	Jul	Aug	Sep	Oct	Nov	Dec	Year
Calm	70	63	82	92	109	122	94	67	87	82	77	32	977
Light Airs	82	70	70	72	106	128	161	115	120	121	103	85	1233
Light Wind	412	321	365	316	430	510	569	590	532	525	434	411	5415
Moderate Breeze	641	635	718	714	597	497	488	502	478	502	510	577	6859
Fresh Breeze	177	152	169	143	165	177	76	67	71	86	139	128	1550
Fresh Gale	69	66	87	50	59	69	26	14	33	47	42	39	601
Strong Gale	6	6	1	5	3	8	1	10	7	3	3	5	58
Hard Gale	0	0	0	0	0	0	0	0	4	2	0	0	6

TABLE 9

Wind force scales used by different observers in the 18th and early 19th centuries. The final column provides equivalent wind speed values for the Beaufort Scale only. The other wind force terms cannot be directly compared with this final column of wind speed values. Asterik indicates that "Windy" is interchangeable with "Fresh Breeze" in the Charleston column. Data Sources: HCS Portfield, British Library, India and Oriental Office Collection, L/MAR/B/609A. Charleston, SC weather data from various issues of the South Carolina Gazette. Beaufort Scale (Smithsonian Institution, 1984). Defoe (1704), as quoted by Lamb and Frydendahl (1991). Thistlewood served as a sailor on the Portfield (L/MAR/B/609E) and his exposure to wind force terms used on this voyage does not appear to have affected his subsequent usage in his own weather journal.

English mariner usage about 1700 (Defoe, 1704)	Terms used in HCS Portfield Logbook, 1746-1748	Terms used in Charleston, SC diary, 1759-61	Terms used by Thistlewood 1750-1786	Royal Navy usage in 1810s (Chenoweth, 1998)	The Beaufort Scale, modern usage	Equivalent speed, knots
Stark calm	Calm	Calm	Calm	Calm	Calm	0
Calm weather	Light airs and calms	Little Wind	Light Airs and calms	Light Airs and calms	Light Air	1 to 3
Little Wind	Light Airs	Small Breeze	Light Airs	Light Airs	Light Breeze	4 to 6
Fine Breeze	Little Wind	Brisk Breeze	Light Wind	Light Wind	Gentle Breeze	7 to 10
Small Gale	Light Gales	Fresh Breeze*	Moderate Breeze	Moderate Breeze	Moderate Breeze	11 to 16
Fresh Gale	Moderate gales	Brisk Gale	Fresh Breeze	Fresh Breeze	Fresh Breeze	17 to 21
Topsail Gale	Fresh Gales	Fresh Gale	Fresh Gales	Strong Breeze	Strong Breeze	22 to 27
Blows Fresh	Prosperous Gales	Stormy	Strong Gales	Fresh Gales	Moderate Gale	28 to 33
Hard Gale of wind			Hard (Heavy) Gales	Strong gales	Fresh Gale	34 to 40
Fret of Wind			Storm	Hard Gales	Strong Gale	41 to 47
Storm			Hurricane	Storm	Whole Gale	48 to 55
Tempest				Hurricane	Storm	56 to 63
					Hurricane	Over 63

Thistlewood was employed by and made one voyage with the English East India Company (EEIC) in 1746–1748 on the Honourable Company's Ship (HCS) *Portfield*. Table 9 includes the wind force terms in the logbook of this ship when sailing from England to India. There is no evidence that Thistlewood derived his wind force terms from that used by the EEIC. Instead, he appears to have used terminology of his own, or in common provenance among nonmariners. Table 9 shows an evolution of wind force terms with the older sense of "gale" becoming two specific terms: (1) breezes for lighter winds, and (2) gales for stronger winds.

Like all mariners of the time, Thistlewood's wind directions are given on a thirty-two-point compass. Most land-based weather records of the eighteenth and nineteenth centuries examined by this author usually employ a sixteen- or eight- point measure of precision of wind direction. When a thirty-two-point scale is used, it is good evidence that the observer is a former mariner, or otherwise aware of their practices.

Table 10 provides the monthly frequency of each observed wind direction for the period 1760–1786. Winds from 240 degrees (WSW) through 350 degrees (N by W) are rarely observed. This is an accurate reflection of Jamaica's location in the Atlantic Trade Wind Zone where easterly winds predominate year round. Westerly winds are rarely observed, and are observed during the winter months when mid-latitude storms reach abnormally far to the south and completely displace the normal high pressure in the region. During the summer rainy season, westerly winds are a sign of tropical depressions, tropical storms and hurricanes.

Thistlewood frequently recorded his wind direction as "variable". He explains his usage:

> What I mean by light Airs, or Winds, Variable, is when the Breeze comes perhaps this instant from the South, in a few Minutes after from the East, and in a little time again from the NE, SE, South or SW, just as it happens, and what I call Mod[erate] Breezes, Variable is much the same, only stronger, commonly continuing longer, perhaps half an hour or an hour the same way; and then at those times, out at Sea, two or three miles from the Land, the Sea Breeze shall blow constant and regular enough.[13]

In other instances, he would list a range of directions in which the wind shifted, for example, from NE to SE to S. In the database, the two wind direction columns list the first and last stated direction.

---

[13] Thistlewood, Monson Collection, 31/89, p. 64.

TABLE 10

Monthly Frequency of Wind Directions in the Thistlewood Weather Record from 1760 to 1786. As expected from a tropical location, winds from the easterly half of the compass are the greatest in number.

WIND DIRECTION IN DEGREES, AND 32-POINT COMPASS TERMS	Month												Year
	Jan	Feb	Mar	Apr	May	Jun	Jul	Aug	Sep	Oct	Nov	Dec	
360, North	60	54	30	19	5	7	6	14	6	32	55	49	337
11, North by East	1	8	3	4	1	0	0	1	1	2	11	12	44
22, North-Northeast	58	49	23	20	1	2	5	18	8	34	67	90	375
33, Northeast by North	2	1	0	0	0	0	0	1	2	1	2	1	10
45, Northeast	163	85	63	54	20	7	27	47	31	73	179	215	964
56, Northeast by East	9	4	3	0	1	0	0	0	0	0	4	9	30
67, East-Northeast	37	17	17	6	3	4	4	9	10	14	24	27	172
78, East by North	2	0	1	0	0	0	0	1	0	0	2	2	8
90, East	27	17	21	25	14	10	13	9	8	16	22	10	192
101, East by South	0	0	0	0	0	1	0	1	1	1	0	0	3
112, East-Southeast	0	1	0	0	4	3	4	5	2	1	2	0	22
123, Southeast by East	0	0	0	0	0	1	0	0	0	0	0	0	1
135, Southeast	30	57	71	64	85	129	88	95	87	85	21	36	848
146, Southeast by South	4	0	0	0	2	5	0	0	1	3	1	0	16
157, South-Southeast	24	37	38	50	69	91	23	48	57	55	13	15	520
168, South by East	10	8	8	9	26	31	10	5	9	18	0	1	135
180, South	48	76	132	152	130	102	78	79	111	103	46	18	1075
191, South by West	0	0	1	1	3	2	0	1	2	2	0	0	12
202, South-Southwest	3	2	11	13	19	8	12	11	3	4	1	1	88
213, Southwest by South	0	0	0	0	0	0	0	0	0	0	0	0	0
225, Southwest	5	6	7	12	13	7	18	6	20	17	1	6	118
236, Southwest by West	0	0	0	0	0	0	0	0	0	0	0	0	0
247, West-Southwest	0	0	0	0	0	0	0	0	0	0	0	0	0
258, West by South	0	0	0	0	0	0	0	0	0	0	0	0	0
270, West	1	0	0	0	0	0	1	0	3	2	1	0	8
281, West by North	0	0	0	0	0	0	0	0	0	0	0	0	0
292, West-Northwest	0	0	0	0	0	0	0	0	0	0	0	0	0
303, Northwest by West	0	0	0	0	0	0	0	0	0	0	0	0	0
315, Northwest	1	1	2	0	0	0	0	0	1	1	1	0	7
326, Northwest by North	0	0	0	0	0	0	0	0	0	0	0	0	0
337, North-Northwest	2	2	2	1	1	0	0	0	0	1	2	2	11
348, North by West	0	7	1	0	0	0	0	0	0	1	1	1	12
0, Calm	6	11	4	12	39	31	30	18	24	28	18	2	223

The sea and land breezes are fundamental aspects of the coastal Jamaican climate. At night, land breezes flow out toward the warmer offshore waters, and during the day the cooler sea breezes flow into the heated land surface. At Savanna-la-Mar, there is no modern climatology available. Thistlewood's data indicate that throughout the year, on average, a light northeast wind prevails at night while during the middle of the day, a south to south-southeast sea breeze dominates. During the months of November through February, the southerly daytime sea breeze is weaker due to the presence of high pressure to the north of Jamaica, and is frequently absent. The remainder of the year, with high pressure normally centered to the east and northeast of Jamaica in the central Atlantic, the daytime sea breeze is normally present. The description of both plantations given in Appendix 1 indicates that both sites provided a representative flow of air and that his determination of wind direction and force were not subject to local site peculiarities.

Because most wind force observations were made during daylight hours, the best record of wind data underestimate the nighttime land breeze. Within daylight hours, winds range from the light land breezes soon after sunrise, to the full strength of the sea breeze in the midday and early afternoon. Table 11A lists wind force terms by the daytime TODs. The purpose is twofold.

First, it identifies which TOD provides the largest sample size that can best represent the actual prevailing wind force. Second, it depicts the diurnal rise and fall of wind speed. Table 11B converts the values from Table 11A into the percentage of all observed wind forces. Italicized values in Table 11B show the most common wind force for each TOD. The evening observations have the highest percentage of calm winds for any of the TODs. This shows that the sea breeze has on average diminished by this time. Fresh breezes are most common in the "middle of the day" (MOD) and "P.M." TODs, indicating the peak hours of the sea breeze. Moderate breezes are most common in the "A.M." TOD and the two other TODs (">mn" and ">am") indicating the next time period after "morning" and "A.M." time periods, respectively. This indicates that the usage is equivalent for the three TODs and refers to the mid-morning to midday establishment of the sea breeze. Along with this is the "Day" TOD, which shows a maximum of moderate breezes.

Since the "A.M." and "Day" TODs represent the two largest number of wind force observations (11,717), they are used with the >mn and >am TODs to best sample wind force at times that are most comparable year round. This removes any bias that would otherwise occur due to diurnal variations in wind speed.

TABLE 11

Frequency of Thistlewood's wind force Terms by Time of Day (TOD) Descriptor. A: Actual Values B: Percentage of total amount for each TOD category.
LA/LC = light airs/light airs inclinable to calm
LW/LB = light wind light breeze
MB = Moderate Breeze
FB = Fresh Breeze
FG = Fresh Gales
SG = Strong Gales
MN = Morning; >MN = first observation (exact time unstated) following a stated "Morning" observation; >AM = first observation (exact time unstated) following a stated "AM" observation; MOD = Middle of the day; PM = Afternoon; EVE = evening; DAY = daylight hours, sunrise to sunset

A

	MN	>MN	>AM	AM	MOD	NOON	PM	EVE	DAY	SUM
Calm	220	3	3	191	3	9	73	391	26	919
LA/LC	117	6	7	640	4	5	173	131	115	1198
LW/LB	316	45	45	3176	36	11	524	339	854	5346
MB	52	179	83	3728	226	11	331	206	1785	6601
FB	8	54	12	328	283	16	1730	26	549	3006
FG	2	11	6	60	123	7	60	11	161	441
SG	0	0	0	3	3	2	9	0	11	28
SUM	715	298	156	8126	678	61	2900	1104	3591	

B

	MN	>MN	>AM	AM	MOD	NOON	PM	EVE	DAY
Calm	0.31	0.01	0.02	0.02	0.00	0.15	0.03	0.35	0.01
LA/LC	0.16	0.02	0.04	0.08	0.01	0.08	0.06	0.12	0.03
LW/LB	0.44	0.15	0.29	0.39	0.05	0.18	0.18	0.31	0.24
MB	0.07	0.60	0.53	0.46	0.33	0.18	0.11	0.19	0.50
FB	0.01	0.18	0.08	0.04	0.42	0.26	0.60	0.02	0.15
FG	0.00	0.04	0.04	0.01	0.18	0.11	0.02	0.01	0.04
SG	0.00	0.00	0.00	0.00	0.00	0.03	0.00	0.00	0.00
SUM	715	298	156	8126	678	61	2900	1104	3591

Conversion of Wind Force Terms to Wind Speed Equivalents

Beaufort's first wind force scale [1805] incorporated common terms for estimating wind force in use by the Royal Navy and was officially introduced for use by the Royal Navy in 1837. Beaufort introduced two terms (gentle breeze and moderate gale) that were then rarely used in Navy logbooks. Standardization of the code for international use was made at different times from the late nineteenth century to the mid-twentieth century. These subsequent modifications of the original Beaufort scale and the absence of controlled experiments to calibrate subjective estimates with anemometer values prevents a straightforward conversion of the terms to wind speed values (Kinsman 1969).

Wind force terms for a region of the tropical Atlantic for the years 1815–1816 were compiled, and the frequency distribution of the modern values for the areas was used on the 1815–1816 data. The results indicated significant disagreement (Chenoweth 1999). In short, a "moderate breeze" in the historical data does not convert to the wind speed values for what is called a "moderate breeze" in the modern Beaufort scale.

No local climatology for western Jamaica is available, so ship data from surrounding sea areas was used to estimate the average wind speed at Savanna-la-Mar. Ship data from four areas of the Caribbean centered at 26°N 85°W, 29°N 78°W, 13°N 78°W, and 16°N 69°W were used (U.S. Navy 1974). Each area was assigned a label of a through d, respectively. Since the areas are not equidistant from Savannna-la-Mar, I weighted the areas to approximate a location near Savanna-la-Mar from the equation:

$$(a \times 2) + (b \times 2) + (c \times 7) + (d \times 3) / 14$$

The wind speeds were calculated for Beaufort scale range in the source data (U.S. Navy 1974) (Table 12A). The number of observations for each Beaufort category was converted to the percentile value for each month. These values are given in Table 12B.

The annual values (bottom row of Table 12B) in the modern data are compared to the Thistlewood wind force category percentiles in Table 12C. The cumulative percentile total shows that the wind force categories are not equivalent. For example, in the modern data the two lowest wind-speed categories account for 11.9% of all observations (calms, light airs, and light breezes). Similarly defined wind force terms in the Thistlewood record account for 42.0% of all observations.

The percentile distribution of wind speed values in the modern record were used and applied directly to the Thistlewood wind force percentile distribution. This equates the Thistlewood wind force terms to wind speed at sea near Savanna-la-Mar (Table 12C). However, since the data are not independent of one another, this eliminates the ability to determine long-term changes in wind speed that may have occurred between the eighteenth and twentieth centuries.

An attempt to adjust the wind force terms for frictional effects and diurnal bias in the data gave quantified estimates of wind speed that were at complete variance with the frequency of wind force terms. The reason(s) for this appears to be related to the procedures used by Thistlewood. He apparently used observations of the movement of Trade Wind cumulus and ships at sea, at least in part, to calculate wind direction and speed. His wind force terms are best interpreted as a daily average, with certain exceptions such as detailed descriptions of hurricanes and severe local thunderstorm gusts. For daily average wind speed values, the following wind speeds (in knots) were equated to the wind force terms used by Thistlewood, where Calm = 0 knots; Light airs (and any modifiers used with it) = 3 knots; Light wind or breeze = 8 knots; Moderate Breeze = 13 knots; Fresh Breeze = 18 knots; Fresh Gale = 23 knots; Strong Gale = 28 knots. Monthly scalar and u- and v-component winds were calculated from these values. As with Table 12C, the data are not independent and cannot be directly compared to modern observations.

Miscellaneous Weather Phenomena

Various types of atmospheric phenomena, such as fog and haze, thunder and lightning, waterspouts, meteors, earthquakes, and other natural events (see Appendix 3 for the complete list) are regularly included in the weather journals. The descriptions are normally brief, but give precise times and event duration. Drawings of several waterspouts are included in his journals. He was particularly observant of thunder and lightning. The average annual number of days with thunder or thunder with lightning from 1752–1786 was 190. These numbers are as high as would be expected in a tropical region near a mountain range where thunderstorm activity is virtually a daily event in the rainy season. Definitions of a day with thunder can vary, and Thistlewood's numbers include distant thunder, even if the storm itself did not pass over his station. His inclusion of all thunder heard matches the modern definition of a thunder day in use in the United States.

Thistlewood was aware of distinct differences in the local weather on the island of Jamaica:

TABLE 12

Depiction of method for estimating modern wind speed at Savanna-la-Mar. A. Weighted average centered on Savanna-la-Mar. B. Monthly percentile frequency of wind speeds by Beaufort Scale categories as given in U.S. Navy (1974). C. Modern Beaufort Scale categories and their equivalent wind speeds in knots. The percentile and cumulative percentile values for the modern Beaufort Scale are used to establish that the equivalent wind force terms in the Thistlewood record are not directly equivalent to the modern scale. Numbers do not add up to exactly 100 due to rounding errors.

A

Average wind speed for four areas centered on areas at (a) 26°N 85°W (b) 29°N 78°W (c) 13°N 78°W (d) 16°N 69°W

	a	b	c	d	weighted average
Jan	14.2	14.9	18.0	14.8	16.3
Feb	14.3	15.3	16.9	14.1	15.7
Mar	14.1	14.8	16.2	13.9	15.2
Apr	13.3	13.5	15.3	13.3	14.3
May	11.2	11.5	13.9	14.0	13.2
Jun	9.6	11.2	16.1	15.4	14.3
Jul	8.4	10.8	18.3	15.4	15.2
Aug	8.6	10.5	15.1	14.0	13.3
Sep	11.4	11.5	11.9	12.7	11.9
Oct	13.3	13.2	10.7	11.6	11.6
Nov	13.6	13.7	13.4	11.9	13.2
Dec	13.8	14.0	16.8	13.8	15.3

B

	a 0 to 4	b 4 to 7	c 7 to 11	d 11 to 17	e 17 to 22	f 22 to 28	g 28 to 34	h 34 to 41
Jan	1.4	4.4	15.3	31.5	27.9	14.1	3.7	1.1
Feb	1.8	6.3	17.6	32.1	26.3	11.6	3.7	0.9
Mar	1.8	5.7	18.6	36.1	25.0	10.1	2.4	0.3
Apr	2.8	6.6	20.8	37.6	22.9	7.6	1.6	0.1
May	4.2	8.6	23.1	37.9	20.4	5.4	0.9	<.05
Jun	4.1	7.4	19.7	33.6	24.1	9.6	1.6	<.05
Jul	4.3	7.1	16.3	30.4	26.8	12.9	2.4	0.5
Aug	5.5	10.0	22.1	32.4	21.4	6.6	1.4	<.05
Sep	5.8	12.9	28.5	34.4	14.1	4.0	0.8	0.1
Oct	6.9	14.2	29.4	31.2	13.5	3.9	0.6	0.1
Nov	4.6	9.6	25.0	35.4	18.9	6.3	1.4	0.1
Dec	1.4	5.6	18.0	35.7	25.5	11.4	2.4	0.1
Annual	2.7	8.2	21.2	34.0	22.2	8.6	1.9	0.3

C

	0 to 4	4 to 7	7 to 11	11 to 17	17 to 22	22 to 28	28 to 34	34 to 41	41 to 48
	Calm Light Airs	Light Wind, Breeze	Gentle Breeze	Moderate Breeze	Fresh Breeze	Strong Breeze	Moderate Gale	Fresh Gale	Strong Gale
Modern Ship Data									
Percentile	3.7	8.2	21.2	34.0	22.2	8.6	1.9	0.3	<0.05
Cumulative Percentile	3.7	11.9	33.1	67.1	89.3	97.9	99.8	100.1	100.1
Thistlewood									
Percentile	8.1	33.9	Not used	48.2	7.8	Not used	Not used	2.0	0.1
Cumulative Percentile	8.1	42.0		90.2	98.0			100.0	100.1

The seasons are very different in different parts of this island, in some parishes their dry seasons are in the time of our wet, and their wet or rainy [season] in the time of our dry.[14] Nay, even on the south side of the island, where the seasons in general are the same as ours, the parish of St. Elizabeth is sometimes very wet, when we are dry, but often very dry when we are too wet. In some parts of this island there is very little thunder. I have been informed by people of undoubted veracity, that towards the east end of the parish of St. James, they have indeed, very little in comparison to what we have. For now we have so much that some people take no notice of it, a remarkable instance of what I once saw many years ago. September 21ˢᵗ, 1753, I was with Mr. William Dorrill (who was my employer at that time), assisting him to adjust the bounds of some lands he had upon Prater's Mountains, on the west side of Bluefields. We dined at a tavern a little way up the mountain, kept by one Coaker, where several gentlemen were come to meet Mr. Dorrill, and after dinner one of them named Mr. Tho. Torrent, lay down to sleep, according to the Creole custom, when came on a shower of rain, with thunder and lightning. The lightning struck the north stable not twenty yards from the house and killed a horse belonging to another gentleman, John James Gorse; that notwithstanding the terrible flashes of lightning, and the surprising loudness of the thunder (for every body thought the house had been coming down upon them), Mr. Torrent was not in the least disturbed, and when he awoke after the shower was over knew nothing of what had happened . . . .[15]

The different perceptions make it impossible to quantify the relative differences in thunderstorm climatology over the island.

Temperature

Thistlewood provides considerable information about his temperature records. Thistlewood describes his thermometer and observing procedures:

My Thermometer is supposed to be a pretty good one, was made by Bury'n Martin according to Fahrenheit's scale . . . I have been thus particular to prevent mistakes, as there are thermometers of so many various constructions. Nay! Even Fahrenheit's is sometimes differently numbered, for the Revd. Dr. Niles in the last edition of his *Vegetable Staticx* [sic] and the ingenious Mr. Stillingstreet in his *Tracts*, instead of 32, put 0 at the Freezing Point, and number each

[14] The western portions of Jamaica's northern coast as well as the southern coastal areas near and west of the capital, Kingston, receive only 35 to 40 inches of rain annually and are the driest areas of the country. The mountains in the extreme eastern section of Jamaica receive from 200 to 300 inches per year.

[15] Thistlewood, Monson Collection, 31/89, pp. 59–62.

way, making those below 0, –1, –2, & which as he observes, is indeed more simple, natural and uniform.

My Thermometer hangs about 6 feet from the ground and 15 or 16 feet above the surface of the Sea at high water mark, against a earthen wall, upon which the Sun never shines, but to which there is a free access of air; nor are our Rooms heated by Fire as I do not remember to have seen but in one dwelling house since I came to the island, that had a chimney. . . .

. . . .I have observed that the Thermometer is commonly higher by a degree and sometimes a degree and half, or Two degrees, by one, or two o'Clock in the afternoon, than it was at Noon. And that it keeps gradually falling all night, being lowest about Sunrise that is in the whole 24 hours. Indeed it would be curious enough to observe its height every hour, but as I have my living to seek, would require more time and attention than I can bestow. In the preceding diary its height is given 3 times, viz. At Sunrise, Noon, and Sunset, every day the year throughout, except when it is dotted, at which times was not at home, or employed about something that would not permit my attendance. . . .

When the Thermometer is at 79 or 80, and calm, the heat affects us more, than when it is at 88 or 90 with a moderate breeze; for we have then a succession of fresh air. We are so relaxed, and the pores of the skin so opened, by constant heat, that I have often been very chilly when the thermometer has not been lower than 76, and the Negroes have complained much of the cold.[16]

In a letter to Edward Long, dated 17 June 1777, Thistlewood adds an important fact about his thermometer exposure practices:

With intention that my account might be as near the mark as possible, & the present position of my Therm[omete]r, gives the Noon & Evening heat of the external air, the best of any where I have to place it, but every Evening, move it to under the eaves on the outside of my house, where it hangs till I have noted its height at Sunrise, & then returned to its station. [I] forget, whither, I mentioned this removal in the papers sent you last year; tho't it necessary. Have a pocket one also, by which find the difference between the internal & external air, in a morning, is commonly 4 or 5 degrees & sometimes more, & have lately read with great pleasure your judicious remarks on that head, in the history of Jamaica. A thermometer suspended at a considerable distance from the roof, in a large Cattle Mill house, should imagine, would be an excellent position, but [I] have no such convenience.[17]

Thistlewood's attention to thermometer exposure is commendable. In fact, his thermometer exposure practices are superior to

---

[16] Long Collection, Add. Ms. 18275B, ff. 64.
[17] Long Collection, Add. Ms. 18275A, ff. 122–124.

most of his contemporaries and many subsequent observers down through the end of the nineteenth century. First, he obtained thermometers of good manufacture. Second, he was aware from his own experience and his scientific readings of the unreliability of many thermometers. Third, he gives an exact altitude of his thermometer above the sea and placed it in a shaded location (although reflected radiation can still reach his thermometer, in the absence of a thermometer screen, not invented until 1864, his exposure is the best he could find). Fourth, he exposed the thermometer so that it had as free a flow of air around as possible. His comment on a "large Cattle Mill" being suitable suggests that free circulation of air, the elimination of solar radiation from access to the thermometer, and a near-ground location demonstrate his careful observations and knowledge of the factors affecting the temperature indicated by a thermometer. Fifth, he was aware of, and avoided the risk of, artificial heat input from fires that would increase the temperature above that of the free air. Sixth, he documents his daily movement of the thermometer each night, an important fact that requires the temperature data to be placed into two subsets for analysis. Finally, he was aware of the "wind chill" effect of a breeze and that human perception of heat and cold can be at variance with the temperature indicated by the thermometer.

It cannot be emphasized enough that this type of information is often totally lacking from many historical weather records. This is true even of records of good quality. Thistlewood's close observations are of uncommon caliber for his time. For example, a weather record kept in Salem, Massachusetts by E.A. Holyoke from 1754 to 1829 includes daily temperature data. From 1754 through 1785, Holyoke made his temperatures in an indoor location and only in 1786 did he begin to record outdoor temperature readings (Baron 1991). Another record kept at Cambridge, Massachusetts by John Winthrop from 1742 to 1779 (but with significant gaps) is yet to be properly evaluated. Other observers in England placed their thermometers indoors, frequently in rooms with fires, so that their data are of limited usefulness. Part of the problem was that the instructions given by James Jurin in his 1722 treatise on weather observations were highly influential, but recommended indoor thermometer exposures (Manley 1974). However, Thistlewood read carefully, conducted his own experiments, and learned the best practices independently and much earlier than other contemporary observers.

The monthly average inter-hourly temperature differences are given in Table 13 for both sites at Egypt (January 1764–August

1767) and Bread Nut Island (September 1767–May 1776). The modern inter-hourly temperature range estimated from the San Juan, Puerto Rico hourly temperature data is included for comparison.

From Table 13, it can be seen that noon minus sunset temperature differences are consistently different between Egypt and Bread Nut Island. Changes in thermometer exposure account for this difference, as there was no change in the thermometer throughout the period. When both sets of differences are compared to the modern record (from a standard exposure), then the largest differences involve the two sets using the sunset observation time. By comparison, the noon minus sunrise difference is nearer the modern values. This suggests that the thermometer exposure was more sensitive to incoming radiation during the late afternoon hours than at either noon or sunrise. It is apparent from the enhanced noon minus sunrise and sunset minus sunrise observations that the Bread Nut Island exposures were less sheltered than the Egypt exposure. However, comparing the noon minus sunrise inter-hourly differences at both sites with the modern values gives the smallest differences compared to the modern values. Since it is assumed that the inter-hourly temperature difference does not significantly change through time, only the sunrise and noon observations are used to calculate average monthly temperature.

A comparison of the wet season (May–October) and dry season (November–April) noon minus sunrise differences indicate that the Breadnut Island dry season difference is statistically significant at the .01 level with the modern data. But in the wet season, there is no statistically significant difference. At Egypt, the wet season noon minus sunrise temperature difference is statistically significant at the .01 level. In the dry season, the difference from the modern noon minus sunrise temperature difference is not statistically significant.

CONTEMPORARY DESCRIPTIONS OF THE JAMAICAN CLIMATE

During Thistlewood's residence in Jamaica, several other observers made weather observations in the Kingston area. No other records in the Savanna-la-Mar record are known to exist. Long made weather observations at Spanish Town from October 1760 to December 1763.[18] Long's *History of Jamaica* mentions only three

[18] Long Collection, Add. Ms. 18963. Long (1774, p. 674) mentions observations made from 1752–1759 that "were not taken in Spanish Town, but nearly in the same parallel, and just within the South-side mountains." It is possible that Long may have been informed of Thistlewood's record by a third party.

TABLE 13

Noon minus sunrise temperature difference (Fahrenheit) at Egypt Plantation and Breadnut Island. The larger standard deviations at Breadnut Island indicate more reflected radiation reached the thermometer. The early 1764 drought is evident in the enhanced differences, particularly in April and May.

Egypt Plantation

	Jan	Feb	Mar	Apr	May	Jun	Jul	Aug	Sep	Oct	Nov	Dec	Year
1764	10.9	10.1	11.0	11.2	10.0	7.6	7.2	7.9	7.2	5.6	7.7	9.1	
1765	8.6	8.5	9.5	8.4	7.3	7.8	7.7	7.0	7.6	6.8	8.3	10.6	
1766	8.6	9.8	9.7	9.0	7.8	7.6	7.8	7.6	7.4	7.2	8.1	8.9	
1767	8.5	9.7	9.5	8.7	7.2	7.2	8.1	7.8					
Avg.	9.0	9.5	9.9	9.2	7.9	7.5	7.7	7.6	7.4	6.5	8.0	9.5	8.3
St. Dev.	1.04	0.62	0.64	1.14	1.16	0.25	0.38	0.39	0.21	0.81	0.32	0.92	

Breadnut Plantation

	Jan	Feb	Mar	Apr	May	Jun	Jul	Aug	Sep	Oct	Nov	Dec	Year
1767	9.8	12.7	12.1	10.9	9.8	8.8	10.0	10.3	10.0	8.4	9.3	8.6	
1770	14.7	10.8	13.9	11.1	8.5	8.7	8.9	10.1	10.3	9.5	9.6	11.2	
1771	10.7	11.2	10.0	11.3	8.8	9.1	9.3	8.7	9.7	9.1	9.7	10.2	
1772	13.8	13.4	11.2	12.0	9.9	9.3	10.0	9.6	8.8	9.7	8.1	9.9	
1773	11.3	13.5	13.4	12.8	9.2	10.9	10.9	10.1	9.2	9.0	8.7	12.5	
1774	11.2	15.0	13.6	12.7	11.4	10.7	9.5	9.9	10.2	10.1	9.2	11.2	
1775	14.6	15.1	14.5	13.3	11.2	10.3			8.8	8.7	10.2	12.6	
1776		13.1	12.7	12.0	9.8	9.7							
Avg.	12.3	13.1	12.7	12.0	9.8	9.7	9.8	9.8	9.6	9.2	9.3	10.9	10.7
St. Dev.	2.02	1.68	1.63	0.94	1.11	0.93	0.70	0.57	0.64	0.60	0.70	1.45	
Estimated Modern Avg.	8.5	7.8	9.4	10.2	9.6	9.0	8.8	8.2	8.8	8.2	8.1	8.2	8.7

specific instances of abnormal weather that can be checked against Thistlewood, but because they are from different locations on the island, some differences are to be expected. Long writes:

> From October 1768 to May 1770 was the longest and severest drought ever remembered in this island, which particularly affected the South side district; for in some of the North side parishes, as St. James and St. Mary, and at the two extremities, St. Thomas and Portland in the East, Westmoreland and Hanover in the West, there were moderate showers. In Liguanea, most of the canes there were destroyed by it. The like calamity befell Vere and Clarendon. Many cotton-trees (a tree of the largest size) were killed; which is the more extraordinary, as their tap-root descends a prodigious depth below the surface. The grass on the lowland pastures and meadows was entirely burnt up. Wells and rivers lost their water . . . . (Long 1774)

Long's description provides the best comparison with the Thistlewood record. The monthly rainfall totals at Savanna-la-Mar are well below normal in 1768–1770. From June 1769 through May 1770, there were only 108 days with measurable rain, the lowest on record. At Savanna-la-Mar, the drought continued into 1771. That the drought was extensive across the island is supported by the only account of fire and burning vegetation in the entire 1752–1786 period at Savanna-la-Mar recorded in the weather journal. On 31 March 1770, Thistlewood describes a haze or mist "with a disagreeable smell" following several weeks of "excessive drying"[19] weather. From 25 May to 4 June, there was a steady fall of moderate to heavy rain showers, apparently less heavy than fell at Kingston. At Kingston, Long (1774) writes that "exceeding heavy rains" fell and ended the drought.

Long also provides an account of eight days of continued rain at Spanish Town in May 1761 that followed "a series of hot, dry and calm weather" (Long 1774). At Savanna-la-Mar, the same month had 20 days with rain (six days above average), and moderate to heavy rains fell each day from 20 to 28 May (nine consecutive days). There is no evidence of "calm" weather preceding this or other rains in the month, so this probably refers to sometime before May or reflects real differences in winds at each site.

Long also specifies a drought from January to June 1764 along the south side parishes of Jamaica in which "multitudes of cattle perished at the pens in the neighbourhood, and dropped in the roads for want of fodder" (Long 1774). The same six months in

---

[19] Thistlewood, Monson Collection, 31/56.

Savanna-la-Mar in 1764 had the third lowest number of rain days on record (45), exceeded only in 1754 (41) and 1786 (37).

Long's surviving abstract of his daily weather observations notes that the year 1763 had been dry in much of the island. There were complaints in March of dry weather on the north side of the island (the normal dry season, so even drier than usual). On 21 August 1763, Long wrote, "There has been no rain in Sp[anish] Town [for] a considerable time."[20] This led into the culminating drought of 1764 mentioned above.

Finally, from a limited number of rainfall measurements made in Jamaica and other islands of the Caribbean, Long estimated the average rainfall of Jamaica "throughout" to be 65 to 70 inches per year (Long 1774). There is a large measure of uncertainty in such an extrapolation. The 1931–1960 average rainfall for the island of Jamaica ranges from 35 to 120 inches per year (locally up to 250 inches in the Blue Mountains). About 90% of the island receives between 50 and 100 inches, so Long's estimate is not unreasonable.

Contemporary newspapers also provide independent evidence of weather conditions in and near Jamaica. The severe drought of early 1786 was felt in Hispanolia with all provisions short and dear (*Pennsylvania Gazette*, 5 July 1786) The severe drought of 1768–1770 was felt in the Yucatan Peninsula of Mexico. Spanish ships arriving in July and early August 1770 in South Carolina from Campeche reported a thirteen-month drought that was prevalent in the provinces adjacent and near the Gulf of Mexico (*South Carolina Gazette*, 9 August 1770). The next year, drought-ending rains may have fallen, but only to be consumed by locusts. According to reports from coastal Belize, ". . . . there had been a Famine among the Indians in the Back Part of the Country, owing to their Crops having been destroyed by Locusts. By which means Provisions of all kinds were so scarce in the Bay, that the oldest settlers did not remember such a severe scarcity at that place" (*Pennsylvania Gazette*, 24 September 1771). An account from Philadelphia, dated 24 October 1771, reported that the locusts were a foot thick in places in the Bay of Honduras region. "At Ambergrease, it is said, that 17,000 Indians had died for want, and in other parts of the country thousands were dead, and dying; so it was computed that upwards of 80,000 Indians had died with famine when the last accounts came away" (*New York Gazette*, 28 October 1771). In Jamaica, drought in the spring of 1754 was severe according to a letter of 6 May 1754:

---

[20] Long Collection, Add. Ms. 18963.

The difficulties the Inhabitants of this island now labor under are almost insurmountable. We have not had one drop of rain for three months past. For want of which, every Thing is in a manner ruined. The rivers that run from the Mountains are almost dried up. Every day we have the repeated news of this and other Plantation and Sugar work being burnt up. Large Cavities appear in many places on this island and we are not with some apprehensions of an approaching Earthquake. We have made frequent requests for rain . . . .

"Capt. John Jauncey, in 27 days from Port Morant Bay, confirms the particular of the above letter, and says, that such dry times and such a scarcity for many kinds of provisions were hardly ever known in the island of Jamaica" (*Pennsylvania Gazette*, 20 June 1754).

Thistlewood wrote that "The year 1754 was excessive dry throughout the island . . . . a great many fires, or thousands of acres of woodland was burnt down, during this extraordinary drought."[21] Additional newspaper accounts of tropical storms and hurricanes affecting Jamaica are discussed in Chapter 5.

A close examination of Thistlewood's record reveals that the record is of unusually good quality and internally consistent. Evidence for changes of observing procedures are documented and the best interpretation of the data, with respect to modern conventions, is described. Thistlewood's data are consistent with the available independent evidence from elsewhere in Jamaica. The analysis permits high confidence to be placed in the subsequent analysis of the data and its comparison with the modern climate of Jamaica, and is the subject of Chapter 4.

---

[21] Thistlewood, Monson Collection, 31/89, pp. 77–78.

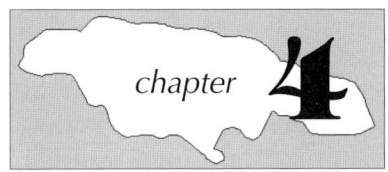

chapter 4

THE EIGHTEENTH CENTURY CLIMATE OF JAMAICA COMPARED TO THE TWENTIETH CENTURY CLIMATE OF JAMAICA

MODERN CLIMATOLOGICAL DATA FOR JAMAICA

Ironically, and sadly, the daily Thistlewood weather record is apparently the most complete weather record for Savanna-la-Mar to this day. A fire at the Jamaican weather center in the early 1990s destroyed most records (Thomas Peterson, personal communication). Monthly rainfall totals and mean temperature are available, but most daily records from which the totals and means are derived are unavailable.

Reliable modern climatologies for Jamaica containing information on all weather variables are few (Jamaican Meteorological Service 1973; Hydrographic Office of the Navy 1993). A comparison of the eighteenth century Savanna-la-Mar record with twentieth century data is also limited because the modern record from this site is a discontinuous series and no metadata are available.

Monthly rainfall totals at Savanna-la-Mar for the period 1920–1977 are available on the Web site at *cdiac.esd.ornl.gov* and were used in this study (Eischeid et al. 1991). The numerous gaps in the record suggest potential discontinuities in the record that indicate possible station relocations and observer changes.

Mean monthly temperature data for Savanna-la-Mar are not available. Instead, the average nighttime marine air temperature (NMAT) for the nearest 1° of latitude by 1° of longitude sea area adjacent to western Jamaica (Bottomley et al. 1990) were used. This method avoids potential urban heat-island biases and station relocation in monthly mean temperature records from neighbor-

51

ing stations such as Negril Point and Montego Bay. The NMAT is adjusted to the mean of 24 hours by using published information on composite diurnal temperature cycles at sea (Parker et al. 1995). The Thistlewood temperature data were adjusted to the mean of 24 hours using hourly temperature data for San Juan, Puerto Rico, which is located at nearly the same latitude as Savanna-la-Mar. This was the nearest easily available station data with complete hourly temperature data.

Other weather elements besides temperature and precipitation are not directly compatible with modern data because of a lack of good modern climatological data for Savanna-la-Mar and also because of different observing methods and procedures. Within-period variability in the Thistlewood record is analyzed and presented.

RESULTS

Temperature

The shortest portion of the Thistlewood record is the temperature series. Although the site location changed in September 1767, the use of different period means for each exposure does not significantly change the fluctuations shown in Figure 2, or the interpretation of the major trends. Figure 2 depicts the seasonal temperature means as anomalies from the entire period of record, 1764–1767, and 1770–1776. Mean monthly temperatures are included in Table 14.

Significant cold anomalies are apparent from summer 1764 through winter 1764–1765, from autumn 1766 through summer 1767, and from winter 1775–1776 into June 1776 (when the record ends). The coldest year was June 1775–May 1776, which was 1.9°F colder than average.

Significant warm anomalies prevailed almost continuously from 1770 to the winter of 1773–1774. This warm period probably began in the spring of 1769. The winter of 1767–1768 produced the coldest January and February temperatures on record, so the cool temperatures in 1767 probably continued into at least early 1768.

The average temperature of 78.60°F from 1764–1776 is 2.80°F lower than the mean of 1951–1980. Assuming a normal distribution of temperatures, 95% confidence intervals were estimated to measure the uncertainty in the estimated average. The estimated error is ±0.39°F. The estimated error from the conversion of the

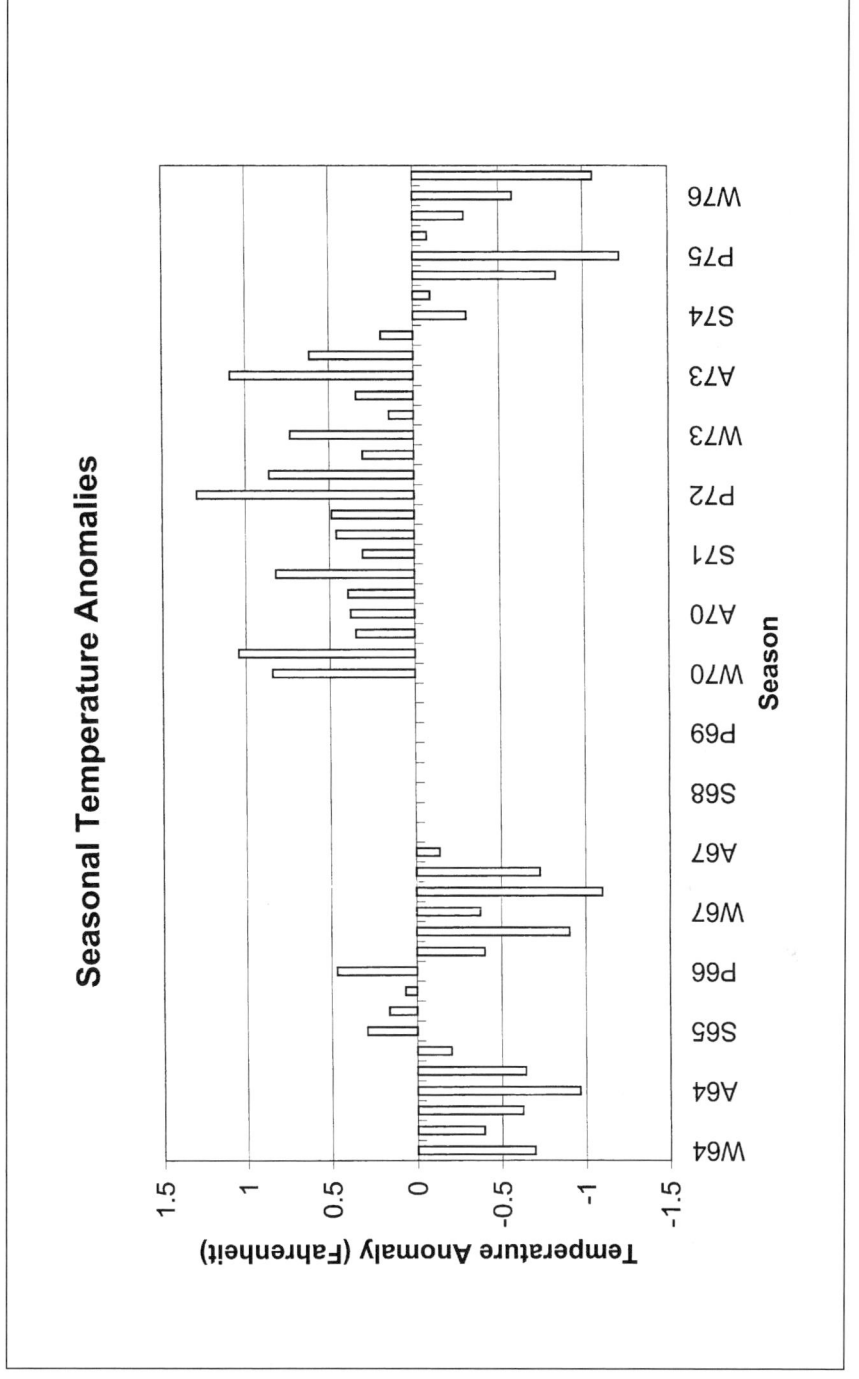

FIG. 2 Seasonal temperature anomalies (Fahrenheit) from 1764 to 1776 at Savanna-la-Mar.

TABLE 14

Monthly and annual mean temperature (Fahrenheit) at Savanna-la-Mar, 1764-1776. Means are the average of 24 hours calculated from the sunrise and noon observations and adjusted for diurnal inequality from hourly temperature data. St. Dev is the standard deviation. The modern reference period 1951–1980 is shown for comparative purposes.

Year	Jan	Feb	Mar	Apr	May	Jun	Jul	Aug	Sep	Oct	Nov	Dec	Annual
1764	75.1	75.4	75.8	77.0	79.3	80.1	79.7	80.2	80.2	78.6	78.3	76.1	77.98
1765	74.6	75.1	76.5	77.1	79.2	80.9	80.7	80.9	81.2	79.6	79.6	76.8	78.52
1766	76.0	75.1	77.1	78.5	79.0	80.5	80.1	80.0	79.7	79.4	78.1	76.8	78.36
1767	75.2	74.6	75.2	76.2	78.9	79.0	80.4	80.3	80.5	79.3	79.8	77.0	78.03
1768													
1769													
1770	76.7	76.8	77.7	78.0	80.1	80.8	81.1	81.0	80.8	80.6	79.7	78.2	79.29
1771	73.8	76.8	76.2	78.7	80.0	81.3	80.4	81.0	81.3	80.8	79.3	77.1	78.89
1772	75.6	76.5	77.6	79.1	80.4	81.4	80.9	82.1	80.7	81.0	79.2	77.6	79.34
1773	76.3	76.0	76.6	77.1	79.5	80.8	80.8	81.3	81.7	81.4	80.2	76.7	79.03
1774	76.3	76.0	76.6	77.1	79.5	80.8	80.8	81.3	81.7	81.4	80.2	76.7	79.03
1775	74.4	73.9	75.0	76.8	78.6	80.3	80.5	80.8	81.1	79.9	78.1	77.4	78.07
1776	74.1	74.4	75.4	76.3	76.8								
Avg	75.3	75.5	76.3	77.4	79.2	80.6	80.5	80.9	80.9	80.2	79.3	77.0	78.60
St. Dev.	0.98	0.99	0.92	0.98	0.97	0.69	0.41	0.62	0.64	0.97	0.82	0.58	0.53
1951–80	79.1	78.6	79.2	80.2	81.6	82.8	83.2	83.9	82.9	83.1	81.8	80.3	81.39
St. Dev.	2.02	1.68	1.63	0.94	1.11	0.93	0.70	0.57	0.64	0.60	0.70	1.45	

sunrise and noon temperature observations to the mean of 24 hours is ±0.40°F. The estimated standard error of the Global Ocean Surface Temperature Atlas (GOSTA) climatology for the Tropics is about ±0.02°F (Bottomley et al. 1990). In reality, the uncertainty is greater since a nonstandard exposure was used. Since Thistlewood took great care to obtain accurate readings, a minimum uncertainty of ±0.2°F is assumed. Together, a minimum uncertainty of ±1.01°F is applicable to the annual average temperature.

The estimated difference from the 1951–1980 average is consistent with a previous estimate for Nassau, Bahamas from 1811–1840 of a –0.72°F ±1.18°F (Chenoweth 1998). Given that this is a point location, the sign of the anomaly is consistent with colder Northern Hemisphere Tropics between 1807 and 1827, which were about 1.1°F lower than 1951–1980 (Chenoweth 2000, 2001). The anomaly is not as large as the –3.6 to –5.4°F anomaly estimated from Puerto Rican corals (Winter et al. 2000), but the –1.77 to –3.79°F range of uncertainty for the Thistlewood record just overlaps the low-end estimate (–3.6 °F) of the proxy data.

Rainfall and Rain Days

Table 15 provides the monthly rainfall totals for the period July 1760 through 15 November 1786 and for the modern record from 1920 through 1977. Figure 3 depicts the annual wet and dry season totals for 1752–1786. The average annual rainfall of 59.63 inches in 1760–1786 is very near the 1920–1977 average. Broadly, the annual rainfall is about equal between the two periods. Given that the eighteenth century was cooler, then the reduced evapotranspiration rates indicates moister conditions in the late 1700s than in the mid-twentieth century.

The seasonal distribution of rainfall reveals that the dry season (November to April) rainfall from 1750–1786 totaled 17.65 inches. This is 29.6% of the annual rainfall. For the wet season (May–October) the rainfall was 41.98 inches, 70.4% of the annual rainfall. For 1920–1977, the percentages for the dry and wet seasons are 27.0% and 73.0%, respectively. This indicates that the dry season rainfall was more plentiful and the wet season rainfall less plentiful than in the mid-twentieth century. However, the differences for both seasons between the eighteenth and twentieth centuries are not statistically significant.

The number of days with rainfall (trace amount or more) is given in Table 16 and also depicted in Figure 4. Similar data for the twentieth century are not available. The period of record for rain days is

TABLE 15

Monthly and annual rainfall totals (inches) at Savanna-la-Mar, 1760-1977. Avg. is the average monthly rainfall and St. Dev is the standard deviation of the monthly rainfall. The wet season covers the months May through October and the dry season is from November through April. The total amount for November 1786 is for the first fifteen days only. The annual total for 1786 is only for the period through 15 November 1786.

Thistlewood Record

Year	Jan	Feb	Mar	Apr	May	Jun	Jul	Aug	Sep	Oct	Nov	Dec	Annual	Wet Season	Dry Season
1760							6.63	7.09	5.94	4.32	1.51	1.53			19.74
1761	1.75	1.19	1.99	11.77	8.62	1.18	4.50	10.24	8.48	7.64	3.52	2.84	63.72	40.66	10.26
1762	0.82	1.19	0.63	1.26	6.97	6.89	4.03	5.86	5.14	3.24	4.79	0.47	41.29	32.13	23.21
1763	5.22	5.92	2.63	4.18	5.06	4.62	5.94	5.58	4.95	8.50	3.07	3.94	59.61	34.65	10.22
1764	0.08	1.56	1.57	0.00	4.89	4.58	6.84	8.95	4.43	14.05	1.56	1.77	50.28	43.74	12.08
1765	1.42	1.88	1.21	4.24	7.33	3.35	3.37	6.74	3.06	5.70	3.15	0.45	41.90	29.55	16.49
1766	5.47	1.16	2.38	3.88	10.70	6.43	6.37	12.45	6.85	6.55	3.39	1.51	67.14	49.35	24.49
1767	5.24	1.74	3.73	8.88	9.31	7.30	5.79	6.97	7.57	8.67	0.61	3.61	69.42	45.61	11.31
1768	0.97	2.39	2.04	1.69	2.82	4.84	2.51	13.22	6.86	5.24	3.83	1.91	48.32	35.49	21.89
1769	1.36	0.43	6.29	8.07	1.65	10.41	4.50	7.30	7.20	2.09	0.59	0.42	50.31	33.15	7.46
1770	1.81	1.46	0.07	3.11	6.80	4.25	5.67	7.76	5.37	5.51	3.85	1.06	46.72	35.36	11.71
1771	0.35	0.39	1.94	4.12	7.32	2.60	6.69	8.34	1.63	4.95	3.06	2.90	44.29	31.53	13.44
1772	1.18	2.62	1.60	2.08	3.94	6.73	9.43	4.84	13.91	3.23	6.23	3.33	59.12	42.08	17.84
1773	0.41	1.67	4.24	1.96	6.90	2.64	7.74	9.44	3.18	6.87	5.76	0.08	50.89	36.77	21.70
1774	3.51	2.48	6.77	3.10	5.41	6.28	10.85	2.58	2.97	7.15	4.08	1.58	56.76	35.24	25.43
1775	4.26	1.30	8.81	5.40	3.55	6.87	7.69	8.53	7.09	14.02	5.69	2.31	75.52	47.75	14.03
1776	0.34	0.85	2.58	2.26	10.11	8.99	2.81	7.73	3.13	10.82	4.01	1.65	55.28	43.59	19.37
1777	1.82	2.44	4.22	5.23	15.12	6.06	8.67	6.66	9.70	7.84	4.67	0.45	72.88	54.05	23.67
1778	0.27	0.41	6.26	11.61	13.71	5.70	6.60	8.40	7.59	6.47	2.06	2.60	71.68	48.47	20.67
1779	1.85	0.54	2.79	10.83	14.58	6.38	4.50	9.93	4.77	3.70	3.49	0.64	64.00	43.86	16.50
1780	3.53	2.04	2.06	4.74	5.40	7.74	4.04	15.10	5.23	37.00	0.60	0.74	88.22	74.51	17.65
1781	3.41	2.27	2.15	8.48	5.73	5.82	5.75	13.16	8.88	3.56	3.17	3.70	66.08	42.90	24.70
1782	1.32	5.39	9.05	2.07	5.01	3.97	7.65	6.75	2.79	1.80	7.65	1.43	54.88	27.97	28.52
1783	7.22	2.05	3.85	6.32	7.18	3.16	6.89	6.26	4.69	7.79	3.00	3.39	61.80	35.97	16.38
1784	0.67	2.80	3.86	2.66	9.40	11.30	15.32	11.89	7.06	4.01	2.53	4.48	75.98	58.98	21.05
1785	4.18	1.06	1.96	6.84	4.60	6.60	5.82	15.84	2.14	11.65	1.89	1.03	63.61	46.65	11.20
1786	0.04	4.73	3.33	0.18	1.37	13.61	7.98	8.21	4.37	10.68	1.19		55.69	46.22	

Modern Record

Year	Jan	Feb	Mar	Apr	May	Jun	Jul	Aug	Sep	Oct	Nov	Dec	Annual	Wet Season	Dry Season
1920	2.56	0.28	2.80	0.32	7.48	14.25	3.94	8.35	5.83	7.21	7.13	1.93	62.08	47.06	35.28
1921	4.49	3.54	9.96	8.23	14.49	7.91	4.33	11.93	6.26	5.28	5.63	1.46	83.51	50.20	21.03
1922	3.31	4.41	5.08	1.14	7.09	6.02	4.49	5.35	10.75	4.65	1.73	3.47	57.49	38.35	14.25
1923	0.67	0.67	2.32	5.39	7.80	1.69	5.39	4.37	8.43	12.01	0.75	0.75	50.24	39.69	

Year															
1924	0.24	1.97	2.09	0.79	8.58	13.11	8.47	4.84	8.31	15.83	2.13	2.24	68.60	59.14	6.59
1925	2.60	5.47	0.32	2.68	2.56	6.58	5.67	7.68	9.21	6.06	3.90	0.75	53.48	37.76	15.44
1926	0.71	1.85	1.73	8.82	7.28	6.65	5.24	12.68	9.06	11.97	2.13	0.63	68.75	52.88	17.76
1927	0.00	1.14	0.95	1.77	14.29	6.10	5.87	4.25	4.25	17.87	3.07	0.24	59.80	52.63	6.62
1928	1.38	0.43	1.10	4.09	5.71	5.04	3.43	8.66	3.70	10.08	3.19	2.01	48.82	36.62	10.31
1929	1.02	0.35	2.48	2.36	13.19	5.24	5.71	12.01	4.80	7.21	1.54	2.95	58.86	48.16	11.41
1930	1.14	6.81	0.83	5.00	7.13	5.08	6.02	7.95	4.53	2.01	1.18	1.85	49.53	32.72	18.27
1931	4.88	0.35	1.30	5.59	5.87	9.53	3.70	2.21	7.13	4.92	1.81	0.87	48.16	33.36	15.15
1932	2.36	0.04	1.54	7.32	7.24	4.25	8.94	8.15	5.87	5.20	6.34	0.20	57.45	39.65	13.94
1933	1.34	0.32	6.14	2.72	5.95	13.66	8.78	8.23	10.67	18.15	7.36	0.67	83.99	65.44	17.06
1934	1.77	6.77	8.35	6.81	9.65	1.58	6.02	6.65	6.97	3.78	4.29	4.88	67.52	34.65	31.73
1935	0.12	0.00	10.63	10.55	14.76	8.35	7.80	9.76	9.09	5.75	0.55	2.48	79.84	55.51	30.47
1936	3.50	0.59	2.28	5.24	7.01	7.56	3.70	8.03	10.24	7.56	5.51	3.70	64.92	44.10	14.64
1941	1.46	2.68	2.21	4.33	4.76	2.95	5.43	9.02	2.87	10.55	5.12	1.38	52.76	35.58	19.89
1942	1.65	2.68	4.53	3.27	12.52	2.13	5.63	9.84	4.29	7.28	0.32	2.48	56.62	41.69	18.63
1943	1.50	1.26	5.04	6.10	9.92	2.64	5.28	10.04	8.50	10.16	4.06	0.43	64.93	46.54	16.70
1944	2.72	0.63	4.45	5.87	0.87	6.85	13.46	4.21	8.78	14.65	2.13	0.20	64.82	48.82	18.16
1945	0.08	3.82	2.56	7.36	4.84	5.75	14.06	12.60	7.91	4.37	3.50	1.77	68.62	49.53	16.15
1946	0.47	1.06	1.85	5.87	7.36	7.44	5.91	5.83	11.30	5.98	2.56	0.00	55.63	43.82	14.52
1947	3.27	0.71	3.19	3.47	9.29	3.03	3.58	3.07	11.69	3.94	0.12	3.07	48.43	34.60	13.20
1948	6.81	0.51	2.32	0.20	9.76	14.25	5.28	2.91	14.84	9.33	3.27	5.35	74.83	56.37	13.03
1949	0.55	0.00	4.96	3.07	9.80	5.20	4.53	7.68	4.84	6.61	3.50	1.93	52.67	38.66	17.20
1950	1.06	0.98	0.89	4.84	5.00	7.44	3.39	2.56	6.42	20.83	2.40	1.89	57.70	45.64	13.20
1951	0.12	3.07	0.71	2.24	4.21	5.55	4.17	11.69	7.01	8.50	5.00	1.50	53.77	41.13	10.43
1954										4.72	1.38				
1955	1.34	1.58	7.09	2.99	4.09	9.17	5.04	3.43	3.47	16.69	4.21	3.19	70.58	53.66	21.82
1956	0.63	0.55	4.17	5.98	12.20	5.47	3.94	8.78	6.58	5.67	1.85	1.38	61.71	40.13	
1957	2.09	8.43	0.87	4.84	8.35	7.17	8.03	3.86	7.05			3.50			
1958	5.24	0.71	0.95	1.10	5.98							1.97			
1962	0.32		0.28	7.44	3.27	7.99	8.11	6.34	5.59	11.93	2.05	0.35	53.32	43.23	
1963			1.77	1.61	9.84	7.95	9.13	5.87	6.38	5.47	2.99	0.32			
1964	1.85	1.65	0.63	13.39	5.28	7.95	10.98	5.87	6.38	5.47	3.15		62.95	41.93	22.48
1965	0.32			1.89	6.42						4.57				
1966					8.47										
1967	3.27		1.10	2.40	4.88						0.55		38.78	27.87	
1968															
1975	0.63	1.34	0.32	2.21	3.50	11.61	7.99	9.45	7.09	5.59	2.68	0.32	52.73	45.23	
1976	1.02	2.40	0.87	0.67	6.42	2.44	2.95	3.94	6.34	2.76	5.43	3.07	38.31	24.85	7.96
1977	1.81	4.37	0.00	4.70	7.21	6.22	11.61	9.41	7.36	6.18	0.75	5.35	64.97	47.99	19.38
1760–1786															
Avg.	2.25	2.00	3.39	4.81	7.06	6.09	6.47	8.73	5.74	7.89	3.29	1.92	59.63	41.98	17.65
St. Dev.	2.00	1.43	2.33	3.39	3.63	2.80	2.65	3.15	2.69	6.69	1.78	1.27	11.87	10.05	5.82
1920–1977															
Avg	1.80	2.07	2.84	4.37	7.57	6.67	6.43	7.10	7.19	8.45	3.07	1.91	59.46	43.40	16.06
St. Dev.	1.59	2.13	2.65	2.92	3.28	3.34	2.79	3.14	2.69	4.73	1.90	1.46	10.83	8.79	6.63

TABLE 16

Number of rain days at Black River (1750–1751) and Savanna-la-Mar (1752–1786). Wet season totals are for the months of May through October. Dry season totals are for the months of November through April. There were no observations made from 1-10 January 1752 and after 15 November 1786. The average (Avg.) and standard deviation (St. Dev.) is for the years 1752–1786. Black River likely has a different rain day climatology. These totals are for all amounts of rain, trace (<.005 inch) and higher. Because of procedural changes in July 1760, the measurable amounts cannot be discerned from the measurable amounts before July 1760.

Year	Month												Annual	Wet Season	Dry Season
	Jan	Feb	Mar	Apr	May	Jun	Jul	Aug	Sep	Oct	Nov	Dec			
1750									13	25	14	11			
1751	4	5	11	7	11	16	26	23	25	22	17	15	223	133	52
1752	8	18	13	19	21	16	19	27	19	20	13	11	215	128	58
1753	9	9	22	23	26	17	25	26	22	15	8	10	174	122	95
1754	6	11	13	4	13	21	22	24	22	22	10	19	203	124	58
1755	11	13	10	16	18	16	18	22	23	21	12	9	174	120	68
1756	10	4	7	12	15	21	29	26	26	27	16	11	256	155	62
1757	14	20	20	20	23	24	21	21	23	19	15	11	195	125	95
1758	11	13	8	12	19	22	20	21	15	24	13	11	204	118	71
1759	13	16	14	19	16	22	21	25	20	14	9	9	163	109	88
1760	6	8	12	10	15	14	21	25	15	18	15	8	176	106	60
1761	8	9	12	18	20	7	20	17	14	16	11	4	154	100	65
1762	10	11	6	12	19	14	25	20	20	19	14	13	206	122	62
1763	8	17	15	17	19	19	28	22	25	20	9	7	170	127	72
1764	3	10	13	1	16	16	16	20	15	14	7	7	158	96	54
1765	10	14	7	14	16	15	21	17	27	15	10	6	182	120	61
1766	16	6	13	13	22	18	17	18	20	22	8	15	199	118	65
1767	9	10	11	20	15	26	12	18	15	18	16	11	167	98	64
1768	7	12	20	8	21	14	12	18	15	18	20	11	170	103	69
1769	14	6	13	17	7	21	15	19	19	22	8	9	167	113	81
1770	10	5	3	13	15	25	17	22	18	16	14	9	167	113	48

(continued)

Table 16 (continued)

Year	Jan	Feb	Mar	Apr	May	Jun	Jul	Aug	Sep	Oct	Nov	Dec	Annual	Wet Season	Dry Season
						Month									
1771	5	7	10	16	19	18	24	21	13	17	12	10	172	112	61
1772	9	12	7	8	17	8	27	21	18	20	19	21	187	111	58
1773	9	14	13	7	17	17	20	18	19	23	18	4	179	114	83
1774	10	11	16	9	17	14	30	23	27	22	13	11	203	133	68
1775	23	13	13	18	22	18	24	24	21	24	12	12	224	133	91
1776	12	14	17	9	21	18	18	25	22	24	13	14	207	128	76
1777	8	8	14	13	23	15	19	19	24	25	15	8	191	125	70
1778	9	6	12	18	20	19	22	20	18	24	14	11	193	123	68
1779	9	12	12	22	23	18	20	21	19	19	13	8	196	120	80
1780	15	10	11	17	13	20	18	23	21	17	7	7	179	112	74
1781	12	9	11	16	15	23	20	24	19	21	20	12	202	122	62
1782	7	20	24	8	16	12	13	25	13	19	15	8	180	98	91
1783	12	14	12	10	15	12	21	23	21	15	9	10	174	107	71
1784	8	12	6	4	18	20	18	28	24	15	6	15	174	123	49
1785	10	5	10	18	18	22	19	26	17	24	11	9	189	126	64
1786	7	15	6	2	15	14	21	19	17	16	10		142	102	50
Avg.	9.9	11.3	11.9	13.2	17.9	17.6	20.8	22.1	19.9	19.7	12.7	10.4	187.1	117.9	68.9
St.Dev.	3.61	4.14	4.43	5.73	3.69	4.4	4.19	2.98	3.9	3.56	3.66	3.68	22.7	12.4	12.7

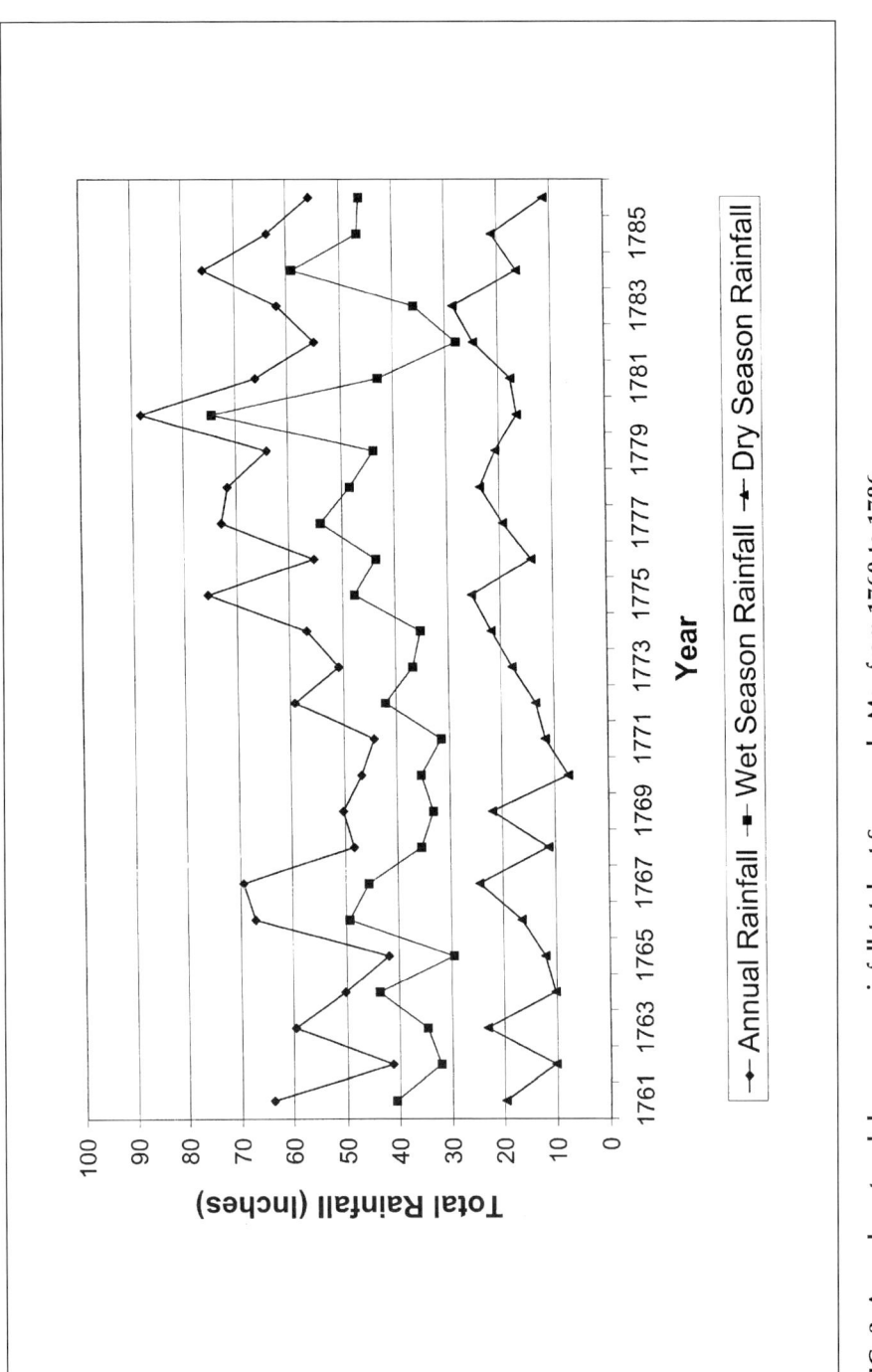

FIG. 3 Annual, wet and dry season rainfall totals at Savanna-la-Mar from 1760 to 1786.

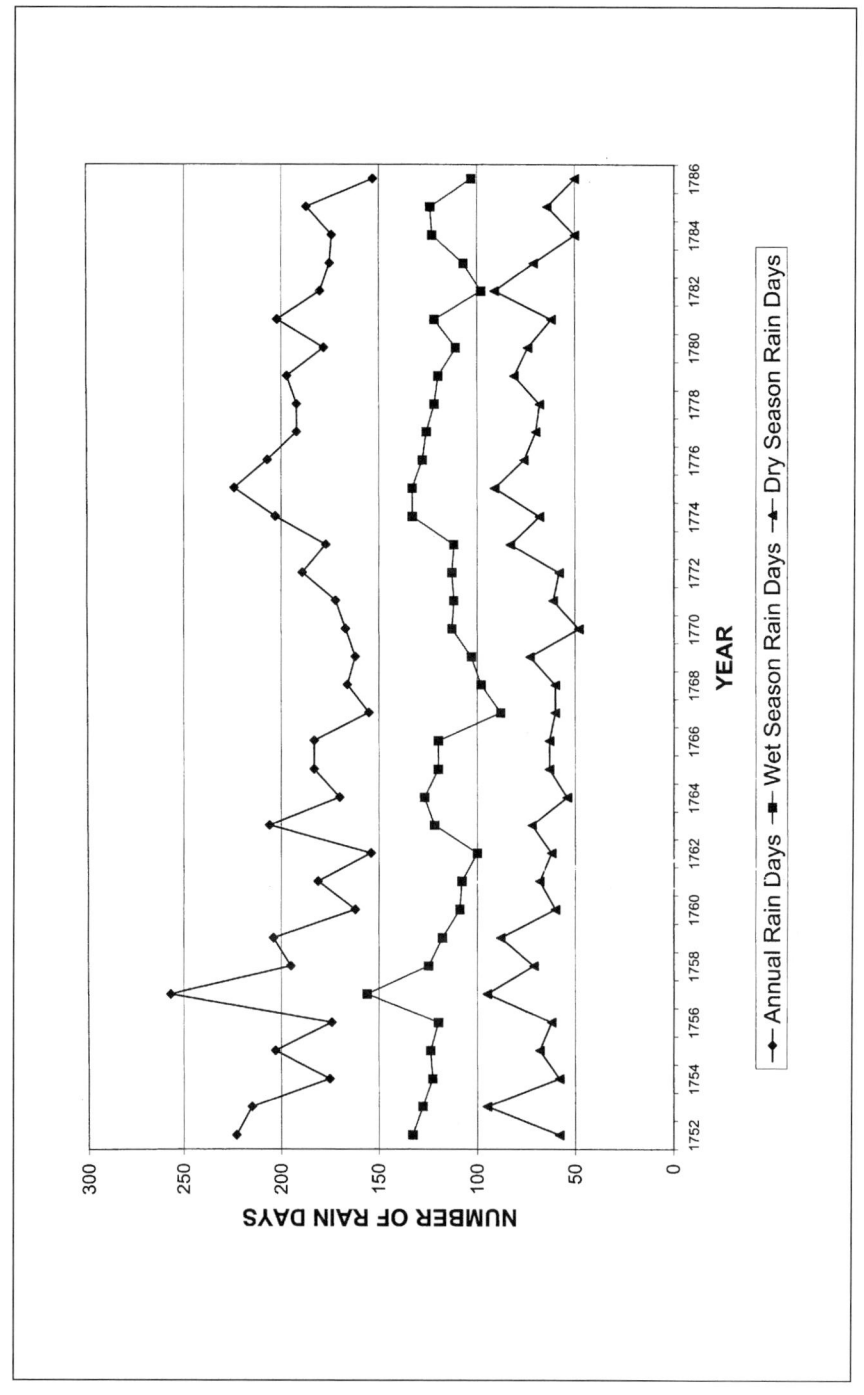

FIG. 4 Number of days with rainfall (annual, wet and dry season) at Savanna-la-Mar from 1752 to 1786.

1752–1786, eight years longer than the measured rainfall record. Because there is evidence that the rainfall descriptors changed relative to the corresponding actual rainfall estimate, the monthly and annual rainfall totals cannot be reliably estimated prior to July 1760. The rainfall days are comparable for the entire period.

The average annual number of rain days was 187 with a standard deviation of 23 days. The year 1757 had the most rain days at 256, and 1762 had the fewest at 154 (1786 possibly had fewer days, but the record is incomplete). The most days with rainfall in the wet season were 155 in 1757, and the most days with rain in the dry season were 95 in 1752–1753 and 1756–1757. The fewest days with rain in the wet season were 96 in 1765, and in the dry season 48 days in 1769–1770.

Figure 4 does not show substantial year-to-year variability in rain days because of the inclusion of nonmeasurable (<.005 inch) trace amounts. However, the extended drought of 1768–1773 is apparent.

Correlation coefficients of rain amounts and rain days (≥0.01 inch) in the 1760–1786 period gave much higher positive correlation with dry season rain days and rain totals (+0.634), but a weaker correlation in the wet season (+0.285). The correlation is significant only in the winter months and is largely due to fewer days of extremely high rainfall amounts during the dry season. In the hurricane season, rainfall totals can vary dramatically, which substantially reduces the correlation.

Sunshine Index and Thunderstorm Days

Figure 5 presents the annual and seasonal variations in cloud cover. The drought of the late 1760s and early 1770s is apparent. Cloudier years were from 1757–1759 and generally after 1773. In general, there is similarity with rain day variations. The year 1757 has the highest wet season cloud cover (i.e., lowest sunshine index values) and number of rain days, but this is not the case in the winter of 1756–1757, which has the highest number of rain days along with 1752–1753, but near average cloud cover. The cloudiest dry seasons were 1762–1763, 1778–1779, and 1752–1753. The dry season of 1769–1770 had the fewest rain days and was also the most cloud free.

A correlation of –0.46 was found between the sunshine index and rain amount in the dry season, but there was no correlation in the wet season (–0.004). The correlation of the sunshine index with rain days was very weak in both the dry (+0.09) and wet (+0.07) seasons.

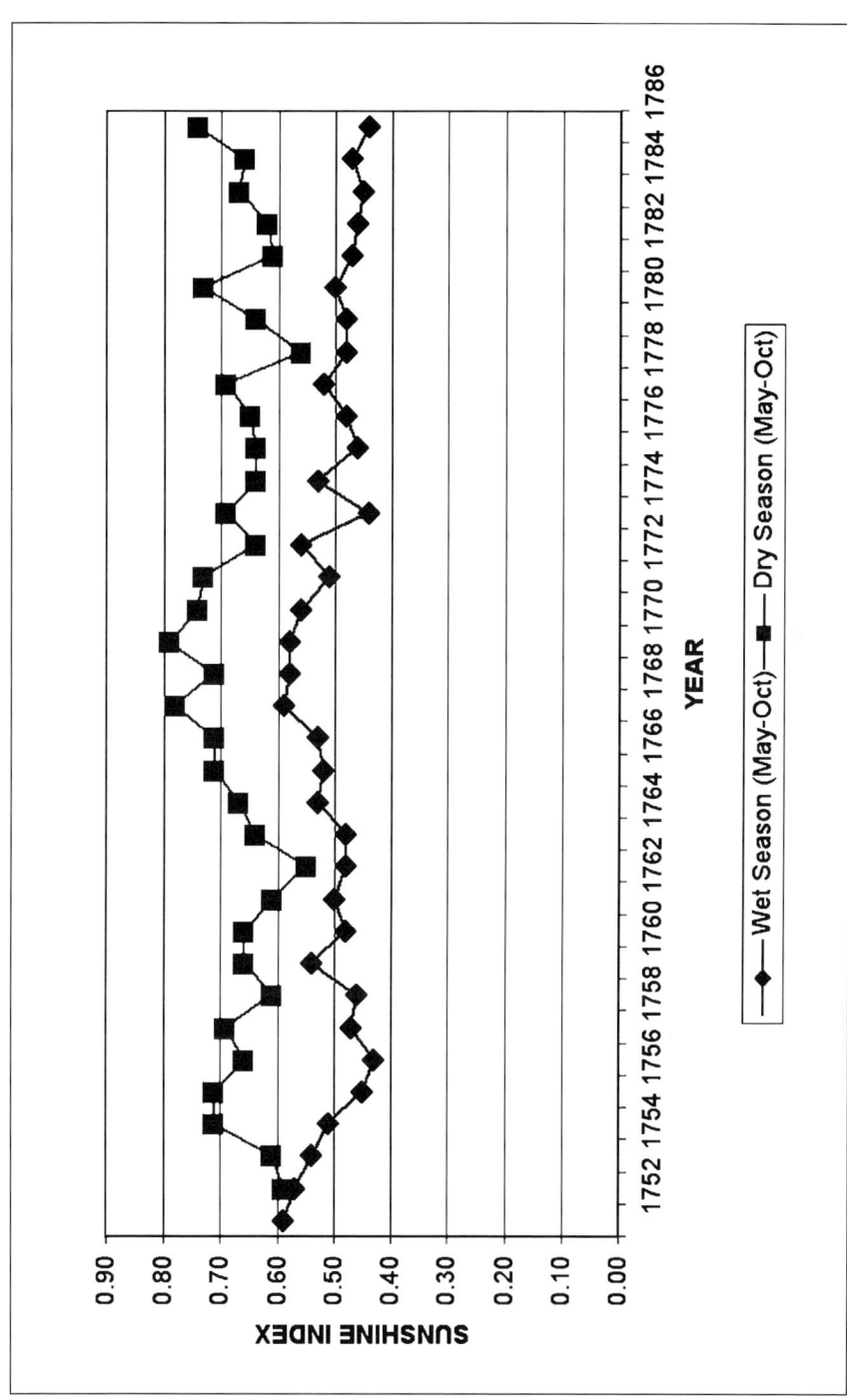

FIG. 5 Dry and wet season sunshine index at Savanna-la-Mar from 1752 to 1786.

Figure 6 presents the annual and seasonal variations in thunderstorm days. Like the cloud cover data, there are no modern data for comparison. However, the sensitivity of Thistlewood is comparable to the thunder day definition in use by the National Weather Service in the United States, where thunder only needs to be heard, regardless of the distance of the thunderstorm from the station.

The wet season has an average of 158 days with thunder. The annual values range from 131 (1754) to 176 (1777). The range for the dry season is from 16 (1754–1755) to 54 (1777–1778). Thunder days were generally below the 35-year average from 1753–1760 (dry season) and 1753–1762 (wet season). Lower numbers returned in the early 1780s in the wet season and the mid-1780s in the dry season. The relatively low number of thunder days occurred during periods of higher than average wind speeds (see next section). Tables 17 and 18 provide the monthly average sunshine index and thunderstorm days, respectively.

Wind

Figure 7 presents annual wet and dry season average wind speed at Savanna-la-Mar for 1752–1786. Table 19 provides the monthly average wind speed values. Wind direction was not routinely recorded until 1760. Windier periods from 1752–1760 and in the 1780s, with lower wind speeds in the 1760s and 1770s. During the dry season, there is an extended minimum in wind speed throughout the 1770s that suddenly rises in the winter of 1778–1779 and remains at higher levels through 1786. During the wet season, the minimum occurs in the first half of the 1770s, and then shows an irregular but very sharp increase to the mid-1780s, when the wind speed was the highest on record.

Figure 8 depicts the zonal (u) and meridional (v) wind components as deviations from the average over the period of record 1760–1786. Table 20 provides the monthly averages of the two wind components. Enhanced easterly (zonal) wind components are more frequent in the 1760s and the end of the 1770s. Enhanced southerly (meridional) wind components are more randomly spread throughout the record.

Figure 9 shows the average monthly wind speed for the period 1752–1786 and compares it with the same months in the modern record (U.S. Navy 1974). Unlike the twentieth century period, there is no secondary wind speed maximum in July. Instead there is a late winter peak in wind speed, followed by a late summer minimum in wind speed.

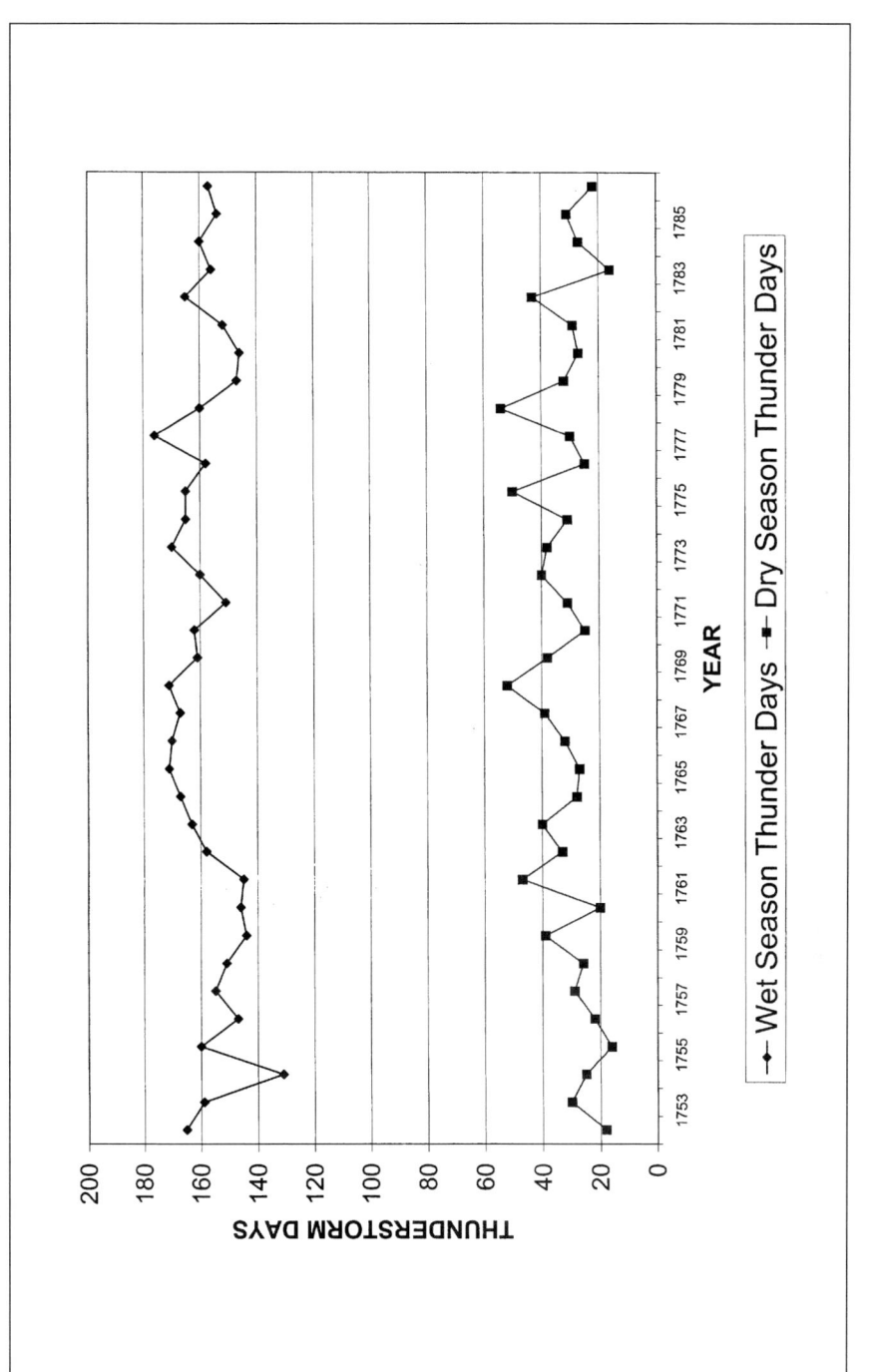

FIG. 6 Number of days with thunder (wet and dry season) at Savanna-la-Mar from 1752 to 1786.

TABLE 17

Average monthly and annual sunshine index at Savanna-la-Mar, 1752 to 1786. Wet season extends from May through October and the dry season extends from November through April. Avg. is the average and St. Dev. Is the standard deviation of the mean. No readings were made from 1–10 January 1752 and after 15 November 1786.

Year	Jan	Feb	Mar	Apr	May	Jun	Jul	Aug	Sep	Oct	Nov	Dec	Annual	Wet Season	Dry Season
						Month									
1752	0.59	0.59	0.71	0.67	0.60	0.67	0.49	0.61	0.47	0.68	0.65	0.77	0.62	0.59	
1753	0.63	0.84	0.31	0.36	0.51	0.55	0.62	0.53	0.53	0.70	0.71	0.87	0.60	0.57	0.59
1754	0.37	0.41	0.57	0.73	0.63	0.42	0.50	0.53	0.55	0.59	0.84	0.76	0.57	0.54	0.61
1755	0.73	0.53	0.81	0.58	0.56	0.42	0.59	0.55	0.55	0.39	0.68	0.67	0.59	0.51	0.71
1756	0.68	0.73	0.77	0.73	0.44	0.36	0.58	0.49	0.32	0.53	0.68	0.79	0.59	0.45	0.71
1757	0.69	0.64	0.53	0.60	0.43	0.44	0.44	0.49	0.34	0.42	0.63	0.72	0.53	0.43	0.66
1758	0.70	0.65	0.77	0.64	0.59	0.49	0.41	0.36	0.46	0.51	0.65	0.63	0.57	0.47	0.69
1759	0.68	0.58	0.62	0.47	0.57	0.42	0.46	0.48	0.49	0.32	0.65	0.67	0.53	0.46	0.61
1760	0.65	0.45	0.85	0.69	0.67	0.65	0.42	0.49	0.41	0.60	0.68	0.71	0.61	0.54	0.66
1761	0.65	0.61	0.64	0.66	0.41	0.55	0.46	0.46	0.62	0.38	0.35	0.53	0.53	0.48	0.66
1762	0.55	0.70	0.90	0.65	0.48	0.54	0.46	0.54	0.53	0.45	0.58	0.73	0.59	0.50	0.61
1763	0.39	0.54	0.53	0.52	0.45	0.44	0.47	0.48	0.45	0.57	0.35	0.59	0.48	0.48	0.55
1764	0.69	0.72	0.64	0.87	0.55	0.51	0.48	0.43	0.44	0.45	0.63	0.77	0.60	0.48	0.64
1765	0.77	0.56	0.66	0.61	0.53	0.53	0.55	0.46	0.61	0.48	0.73	0.87	0.61	0.53	0.67
1766	0.47	0.84	0.69	0.65	0.53	0.61	0.53	0.48	0.49	0.49	0.69	0.82	0.61	0.52	0.71
1767	0.69	0.71	0.74	0.62	0.60	0.47	0.44	0.57	0.57	0.53	0.83	0.68	0.62	0.53	0.71
1768	0.79	0.76	0.84	0.78	0.52	0.62	0.71	0.53	0.68	0.45	0.60	0.85	0.68	0.59	0.78
1769	0.67	0.75	0.72	0.66	0.69	0.57	0.62	0.55	0.57	0.46	0.90	0.84	0.67	0.58	0.71
1770	0.74	0.80	0.89	0.57	0.62	0.54	0.53	0.54	0.52	0.74	0.52	0.90	0.66	0.58	0.79
1771	0.88	0.67	0.79	0.67	0.50	0.59	0.49	0.57	0.60	0.59	0.73	0.85	0.66	0.56	0.74
1772	0.72	0.67	0.72	0.69	0.42	0.68	0.41	0.53	0.48	0.54	0.43	0.58	0.57	0.51	0.73
1773	0.82	0.69	0.65	0.66	0.50	0.53	0.62	0.53	0.60	0.57	0.60	0.85	0.63	0.56	0.64
1774	0.70	0.70	0.59	0.69	0.36	0.56	0.45	0.44	0.35	0.48	0.70	0.70	0.56	0.44	0.69
1775	0.53	0.76	0.53	0.59	0.58	0.66	0.36	0.48	0.61	0.46	0.62	0.74	0.58	0.53	0.64
1776	0.68	0.59	0.57	0.65	0.30	0.49	0.47	0.52	0.53	0.46	0.67	0.57	0.54	0.46	0.64
1777	0.67	0.85	0.50	0.64	0.52	0.54	0.46	0.43	0.53	0.42	0.56	0.88	0.58	0.48	0.65
1778	0.75	0.72	0.66	0.55	0.54	0.48	0.46	0.54	0.50	0.58	0.69	0.60	0.59	0.52	0.69
1779	0.64	0.54	0.54	0.34	0.42	0.45	0.60	0.50	0.52	0.41	0.51	0.79	0.52	0.48	0.56
1780	0.58	0.75	0.65	0.56	0.51	0.35	0.59	0.47	0.52	0.45	0.78	0.86	0.59	0.48	0.64
1781	0.77	0.77	0.69	0.53	0.60	0.49	0.46	0.45	0.45	0.56	0.64	0.65	0.59	0.50	0.73
1782	0.64	0.42	0.48	0.82	0.51	0.33	0.55	0.50	0.48	0.47	0.55	0.80	0.55	0.47	0.61
1783	0.44	0.65	0.75	0.51	0.54	0.45	0.35	0.38	0.49	0.54	0.64	0.58	0.53	0.46	0.62

(continued)

1784	0.67	0.56	0.77	0.78	0.52	0.32	0.42	0.46	0.39	0.58	0.82	0.52	0.57	0.45	0.67
1785	0.64	0.74	0.72	0.54	0.67	0.34	0.46	0.34	0.48	0.51	0.73	0.72	0.57	0.47	0.66
1786	0.81	0.42	0.86	0.89	0.62	0.37	0.43	0.49	0.27	0.47	0.63		0.57	0.44	0.74
Avg.	0.66	0.65	0.68	0.63	0.53	0.50	0.50	0.49	0.50	0.51	0.65	0.73	0.58	0.50	0.67
St. Dev.	0.12	0.12	0.13	0.12	0.09	0.10	0.08	0.06	0.09	0.09	0.12	0.11	0.04	0.05	0.06
Max.	0.88	0.85	0.90	0.89	0.69	0.67	0.71	0.61	0.68	0.74	0.90	0.90	0.68	0.59	0.79
Year	1771	1777	1762	1786	1769	1752	1768	1752	1768	1770	1769	1770	1768	1752	1769–70
Min.	0.37	0.41	0.31	0.34	0.30	0.32	0.35	0.34	0.27	0.32	0.35	0.52	0.48	0.42	0.55
Year	1754	1754	1753	1779	1776	1784	1783	1785	1786	1759	1763	1784	1763	1757	1762–63

TABLE 18

Number of days with thunder heard at Savanna-la-Mar (with or without lightning), 1752-1786. Avg. is the average. An asterisk indicates more than two years had the same value.

Year	Jan	Feb	Mar	Apr	May	Month Jun	Jul	Aug	Sep	Oct	Nov	Dec	Annual	Wet Season	Dry Season
1752	1	5	6	6	26	28	29	31	24	27	9	5	197	165	18
1753	3	0	3	10	23	27	25	28	29	27	11	3	189	159	30
1754	0	4	6	1	11	23	25	29	25	18	1	7	150	131	25
1755	3	0	0	5	20	27	28	31	27	27	9	6	183	160	16
1756	1	0	0	6	26	23	27	29	26	16	7	9	170	147	22
1757	8	1	4	0	18	29	29	29	27	23	6	3	177	155	29
1758	6	0	1	10	21	26	27	28	28	21	17	2	187	151	26
1759	2	3	1	14	22	27	27	26	20	22	6	4	174	144	39
1760	0	1	0	9	20	17	28	29	29	23	6	9	171	146	20
1761	2	2	11	17	23	15	31	30	24	22	14	8	199	145	47
1762	6	0	0	5	24	24	27	31	29	23	14	3	186	158	33
1763	10	2	5	6	24	26	30	28	29	26	11	10	207	163	40
1764	4	0	3	0	22	28	30	31	30	26	10	4	188	167	28
1765	0	6	0	7	23	28	30	31	28	31	14	4	202	171	27
1766	4	2	0	8	30	25	31	30	30	24	14	3	201	170	32
1767	6	2	8	6	23	29	29	30	29	27	22	4	215	167	39
1768	1	11	8	6	26	28	31	28	28	30	19	6	222	171	52
1769	6	0	4	3	18	27	28	29	29	30	12	3	189	161	38
1770	1	1	1	7	20	29	31	31	26	25	16	0	188	162	25
1771	0	0	1	14	21	29	28	29	23	21	12	3	181	151	31
1772	7	3	2	13	25	24	31	30	27	23	19	9	213	160	40

(continued)

Table 18 (continued)

Year	Jan	Feb	Mar	Apr	May	Jun	Jul	Aug	Sep	Oct	Nov	Dec	Annual	Wet Season	Dry Season
						Month									
1773	2	0	4	4	23	30	31	29	28	29	13	1	194	170	38
1774	2	0	5	10	25	22	31	30	30	27	11	13	206	165	31
1775	5	5	11	5	28	24	30	29	30	24	14	2	207	165	50
1776	0	1	4	4	18	27	30	29	27	27	11	5	183	158	25
1777	1	0	2	11	26	30	31	31	30	28	20	4	214	176	30
1778	7	2	12	9	21	21	30	30	30	28	13	8	211	160	54
1779	3	0	3	5	18	28	23	27	29	22	15	2	175	147	32
1780	1	1	0	8	19	29	30	26	26	16	7	1	164	146	27
1781	6	0	9	6	18	30	24	28	26	26	18	10	201	152	29
1782	4	1	7	3	23	28	30	31	29	24	11	1	192	165	43
1783	1	0	0	3	25	21	30	28	29	23	14	10	184	156	16
1784	2	0	0	1	25	28	29	29	28	21	4	4	171	160	27
1785	5	3	6	9	18	29	28	28	25	26	12	8	197	154	31
1786	1	0	1	0	19	26	29	31	28	24	11		170	157	22
Avg.	3.2	1.6	3.7	6.6	22.1	26.1	28.8	29.3	27.5	24.5	12.1	5.2	190.2	158.1	32.2
Max	10	11	12	17	30	30	31	31	30	31	22	13	222	176	54
Year	1763	1768	1778	1761	1766	1761	1779	*	*	1765	1767	1774	1768	1777	1777–78
Min	0	0	0	0	11	15	23	26	20	16	1	0	150	131	16
Year	*	*	*	*	1754	1761	1779	1759, 1780	1759	1756, 1780	1754	1770	1754	1754	1754–55, 1782–83

TABLE 19

Average wind speed at Savanna-la-Mar from 1752-1786 expressed in knots. One knot = 1.15 miles per hour. There are no observations for 1-10 January 1752 and after 15 November 1786. The wet season includes the months of May through October and the dry season the months of November through April. The averages for the months of May through September is for the time of the year that wind speed in the western Caribbean is most highly (positively correlated with sea surface temperature in the equatorial regions of the eastern Pacific Ocean.

Year	Jan	Feb	Mar	Apr	May	Jun	Jul	Aug	Sep	Oct	Nov	Dec	Annual	Wet Season	Dry Season
						Month									
1752	7.6	12.3	12.2	13.7	14.7	8.6	9.5	9.0	12.1	10.6	13.2	14.2	11.5	10.7	10.8
1753	14.4	12.2	12.5	10.7	9.8	11.1	9.3	9.8	10.0	10.0	12.3	11.8	11.2	10.0	12.9
1754	15.9	13.8	13.0	15.2	12.7	11.0	10.7	10.3	10.1	9.5	13.8	11.0	12.2	10.7	13.7
1755	12.1	13.7	14.1	12.6	10.7	8.2	8.7	8.7	7.5	11.5	13.8	8.0	10.8	9.2	12.9

(continued)

Table 19 (continued)

Year	Jan	Feb	Mar	Apr	May	Jun	Jul	Aug	Sep	Oct	Nov	Dec	Annual	Wet Season	Dry Season
						Month									
1756	12.3	11.7	11.5	11.3	10.6	8.2	9.6	11.6	10.5	14.0	13.3	11.5	10.3	11.8	10.3
1757	14.9	12.0	10.9	11.5	13.2	12.6	12.0	10.8	11.9	14.2	15.2	12.8	12.0	13.2	12.0
1758	14.0	16.1	14.0	15.5	13.2	9.7	8.7	10.3	10.7	10.1	12.0	12.3	11.4	14.5	11.5
1759	11.7	11.2	11.3	10.4	10.3	12.3	10.3	10.1	12.8	10.4	12.7	11.3	11.1	11.4	10.7
1760	13.0	13.6	12.1	12.6	13.7	8.2	8.0	9.1	12.3	13.0	12.0	11.6	10.7	12.3	10.3
1761	12.3	11.9	12.3	12.4	9.7	9.1	8.8	9.7	10.8	9.9	9.6	10.6	10.1	12.1	10.0
1762	10.5	11.8	12.3	12.5	10.0	9.6	10.0	8.8	9.9	9.7	15.1	11.0	10.1	10.9	10.2
1763	11.3	11.9	10.8	9.9	10.5	10.6	9.8	9.9	10.1	10.1	11.7	10.5	10.1	11.3	10.1
1764	13.9	12.5	13.2	12.3	11.3	9.9	9.4	9.2	10.3	11.7	10.0	11.4	10.4	12.4	10.4
1765	11.9	12.2	12.2	9.8	10.6	9.4	10.5	10.7	7.4	10.5	11.2	10.7	9.7	11.6	10.2
1766	12.9	15.4	12.7	10.1	11.9	9.3	9.2	9.7	11.5	13.1	12.7	11.6	10.3	12.3	10.1
1767	12.1	12.0	12.1	11.0	10.6	10.7	11.2	10.3	11.0	10.8	10.9	11.2	10.8	12.3	10.8
1768	13.0	13.2	13.5	9.9	9.5	11.3	11.2	8.2	10.0	10.0	11.2	11.1	10.0	12.3	10.0
1769	12.4	13.3	10.7	13.7	9.9	7.5	8.8	9.1	8.3	9.5	10.5	10.4	9.6	11.4	9.8
1770	13.0	13.7	8.7	11.7	10.7	7.0	10.8	7.9	8.9	8.2	9.6	10.3	9.5	11.4	9.6
1771	12.8	13.3	10.3	9.1	8.3	7.5	9.6	8.3	11.0	10.3	11.1	10.3	9.0	11.0	8.6
1772	12.5	12.4	12.6	8.6	12.8	7.2	9.4	9.2	8.4	10.7	9.8	10.4	9.3	11.6	9.5
1773	10.9	12.3	12.1	12.0	9.2	9.1	9.5	10.3	8.2	10.5	12.5	10.6	9.7	11.1	10.0
1774	11.5	11.2	11.2	9.0	11.4	7.3	8.0	8.5	8.2	10.0	11.1	9.9	8.8	11.4	8.9
1775	10.9	10.9	13.0	12.2	11.8	9.1	9.4	8.6	8.9	8.0	10.6	10.3	10.0	11.0	10.2
1776	10.8	11.5	14.5	10.1	10.6	8.3	8.8	10.2	8.3	10.9	7.9	10.2	9.4	11.1	9.6
1777	13.7	12.8	11.9	10.6	10.1	11.2	9.9	10.2	10.1	10.1	10.3	11.0	10.4	11.3	10.4
1778	12.0	11.2	11.0	11.0	11.0	9.6	9.4	11.9	9.3	10.8	11.5	11.0	10.4	11.2	10.6
1779	14.3	13.1	11.2	12.4	9.9	10.7	10.9	10.8	10.6	12.2	10.8	11.4	10.9	11.9	10.9
1780	13.6	10.7	10.9	8.8	10.5	9.1	9.9	10.3	12.5	13.0	14.0	11.4	10.2	12.0	9.7
1781	13.7	12.3	11.6	10.2	9.3	9.0	8.4	9.5	9.2	8.9	10.9	10.5	9.3	12.9	9.3
1782	11.1	12.8	13.5	13.0	12.3	11.3	9.9	9.9	9.6	11.0	11.1	11.6	11.0	11.9	11.3
1783	12.1	13.6	12.8	11.4	12.8	9.2	10.0	11.6	11.7	13.2	12.8	12.0	11.1	12.2	11.0
1784	9.5	12.4	13.7	14.5	9.1	11.4	11.5	11.1	10.3	10.9	13.1	11.8	11.3	12.6	11.6
1785	12.3	15.3	13.5	15.5	11.3	13.3	11.3	12.8	11.9	10.7	12.7	12.8	12.7	13.0	12.9
1786	12.8	13.0	15.2	13.4	12.8	10.2	11.4	11.7	13.1	13.0		12.6	12.1	12.7	11.9
Avg.	12.1	12.7	12.3	11.6	10.8	9.6	9.8	10.0	10.3	11.2	11.6	11.2	10.3	12.0	10.4
St. Dev.	1.59	1.24	1.43	1.85	1.43	1.55	1.02	1.25	1.45	1.68	1.71	0.76	0.89	0.85	0.93

TABLE 20

U-component (A) and V-component (B) winds in knots. One knot = 1.15 miles per hour. The wet season includes the months May through October and the dry season the months November through April. There are no after 15 November 1786. The winter season months of December through March are correlated (not significantly) with sea surface temperature in the eastern equatorial Pacific Ocean. The westerly component (u) is negative year-round in Jamaica, where trade wind easterlies blow year-round. The southerly component (v) dominates from March through October as indicated by the positive values, but are negative in November through February when northerly winds dominate.

A

						Month										
u-component (knots)																
Year	Jan	Feb	Mar	Apr	May	Jun	Jul	Aug	Sep	Oct	Nov	Dec	Annual	Wet Season	Dry Season	Dec - Mar
1760	-7.5	-6.0	-3.5	-2.7	-2.4	-7.0	-3.0	-1.8	-1.8	-3.5	-6.5	-7.0	-4.39	-3.25	-4.93	-5.67
1761	-8.0	-5.5	-2.4	-3.5	-3.0	-2.7	-5.5	-4.5	-1.8	-2.4	-5.0	-5.5	-4.15	-3.32	-5.48	-5.73
1762	-5.0	-6.0	-4.0	-4.0	-2.1	-3.5	-4.5	-3.5	-2.7	-4.5	-5.5	-8.5	-4.48	-3.47	-4.92	-5.13
1763	-8.0	-4.0	-6.5	-3.5	-2.1	-2.4	-5.0	-4.0	-2.4	-6.5	-5.5	-7.5	-4.78	-3.73	-6.00	-6.75
1764	-10.0	-10.5	-8.5	-5.0	-2.4	-3.0	-5.0	-6.0	-2.4	-7.0	-6.5	-7.0	-6.11	-4.30	-7.83	-9.13
1765	-9.0	-8.5	-4.0	-3.0	-2.7	-3.0	-4.5	-6.0	-5.0	-3.5	-6.5	-6.0	-5.14	-4.12	-6.33	-7.13
1766	-6.0	-6.0	-6.0	-2.7	-2.1	-2.7	-3.5	-5.5	-2.7	-6.5	-6.5	-5.5	-4.64	-3.83	-5.53	-6.00
1767	-8.5	-7.5	-7.5	-3.5	-3.0	-3.5	-3.5	-5.5	-2.7	-6.5	-6.5	-5.5	-5.31	-4.12	-6.50	-7.25
1768	-8.5	-6.5	-4.0	-3.0	-3.5	-3.0	-4.5	-4.5	-2.7	-3.0	-4.5	-8.0	-4.64	-3.53	-5.67	-6.13
1769	-8.0	-4.5	-4.5	-1.8	-3.0	-1.5	-3.5	-8.5	-2.4	-5.0	-6.5	-6.5	-4.64	-3.98	-5.22	-6.25
1770	-7.0	-8.5	-6.5	-2.1	-2.4	-2.4	-3.0	-4.5	-2.1	-2.4	-3.5	-5.5	-4.16	-2.80	-6.18	-7.12
1771	-6.5	-1.8	-3.5	-2.1	-2.1	-2.7	-3.0	-5.5	-4.0	-4.5	-6.5	-3.0	-3.77	-3.63	-3.82	-4.33
1772	-7.0	-8.5	-3.0	-2.4	-2.4	-3.5	-3.5	-5.0	-2.7	-4.5	-5.0	-5.5	-4.42	-3.60	-5.07	-5.38
1773	-6.5	-5.0	-7.0	-3.5	-4.0	-4.0	-5.5	-7.0	-5.0	-5.5	-5.5	-8.0	-5.54	-5.17	-5.42	-6.00
1774	-6.5	-8.0	-6.5	-6.0	-2.1	-4.5	-4.5	-5.0	-3.0	-4.5	-4.5	-5.5	-5.05	-3.93	-6.75	-7.25
1775	-4.5	-8.0	-7.5	-2.1	-1.5	-2.4	-4.5	-2.7	-1.8	-4.5	-5.0	-6.5	-4.25	-2.90	-5.35	-6.38
1776	-6.0	-7.5	-4.5	-0.9	-1.8	-0.5	-2.4	-4.0	-2.1	-2.4	-4.5	-4.0	-3.38	-2.20	-5.07	-6.13
1777	-3.0	-2.7	-5.5	-3.0	-1.5	-3.0	-4.0	-8.5	-4.0	-5.0	-5.0	-6.0	-4.27	-4.33	-3.78	-3.80
1778	-8.5	-7.5	-6.5	-3.5	-2.7	-3.5	-4.5	-4.0	-2.1	-4.0	-6.5	-6.5	-4.98	-3.47	-6.17	-7.13
1779	-8.5	-5.0	-6.0	-5.5	-6.5	-5.0	-5.0	-4.0	-5.0	-7.0	-6.5	-7.5	-5.96	-5.42	-6.33	-6.50
1780	-8.0	-6.5	-5.5	-5.0	-3.5	-6.5	-5.5	-6.0	-4.0	-0.9	-5.0	-2.1	-4.88	-4.40	-6.50	-6.88
1781	-8.5	-1.4	-0.9	-3.0	-1.8	-2.4	-3.0	-5.0	-2.1	-3.5	-3.0	-5.5	-3.34	-2.97	-3.48	-3.23
1782	-7.5	-6.0	-6.0	-0.9	0.3	-1.2	-3.5	-5.5	-1.5	-5.5	-7.0	-4.5	-4.07	-2.82	-4.82	-6.25
1783	-3.5	-5.0	-4.0	-4.5	-0.6	-2.7	-2.4	-5.0	-0.6	-4.5	-6.5	-5.5	-3.73	-2.63	-4.75	-4.25
1784	-5.0	-2.4	-3.0	-2.7	-2.1	-2.1	-2.7	-3.5	-2.4	-5.0	-2.7	-7.0	-3.38	-2.97	-4.18	-3.98
1785	-5.5	-2.4	-2.4	-1.2	-1.5	-1.8	-2.1	-3.5	0.5	-2.7	-3.5	-6.5	-2.72	-1.85	-3.53	-4.33
1786	-2.4	-3.2	-0.3	1.7	0.9	-0.3	1.4	-0.9	0.9	0.5			-0.18	0.41	-2.37	-3.10
Avg.	-6.8	-5.7	-4.8	-2.9	-2.3	-3.0	-3.7	-4.8	-2.5	-4.2	-5.4	-6.0	-4.30	-3.40	-5.30	-5.80
St. Dev.	1.92	2.36	2.05	1.60	1.36	1.51	1.44	1.70	1.43	1.81	1.24	1.49	1.14	1.11	1.22	1.42

Table 20 (continued)

B

v-component (knots)

Year	Jan	Feb	Mar	Apr	May	Jun	Jul	Aug	Sep	Oct	Nov	Dec	Annual	Wet Season	Dry Season	Dec - Mar
1760	-4.0	-2.7	5.6	4.9	9.0	8.4	3.4	2.13	6.35	4.75	-4.5	-3.5	2.47	5.66	0.93	-0.38
1760	-4.0	-2.7	5.6	4.9	9.0	8.4	3.4	2.1	6.4	4.8	-4.5	-3.5	2.47	5.66	0.93	-0.38
1761	-3.0	1.2	3.8	5.0	8.8	5.7	4.5	2.0	6.9	4.2	-4.5	-2.1	2.69	5.33	-0.18	-0.40
1762	-2.7	1.2	5.7	4.4	10.8	8.9	1.5	3.8	3.9	0.5	-5.5	-12.0	1.69	4.87	0.33	0.52
1763	-2.4	1.5	1.4	6.8	8.5	7.4	8.0	4.8	8.1	6.9	-4.5	-5.5	3.39	7.25	-1.71	-2.90
1764	-6.0	-3.5	2.2	8.5	9.6	10.5	7.5	5.7	7.9	4.7	-5.5	-4.5	3.08	7.62	-1.47	-3.20
1765	-5.5	-1.5	5.3	2.8	6.7	8.3	7.0	4.8	6.1	3.6	-3.5	-4.0	2.49	6.05	-1.49	-1.56
1766	-5.5	-0.9	10.3	10.6	8.9	10.3	7.2	4.6	7.8	3.2	-8.5	-9.0	3.23	6.97	1.16	-0.03
1767	-5.5	-0.9	8.1	9.7	12.3	5.8	5.8	7.9	7.7	7.2	-4.5	-6.5	3.91	7.76	-1.03	-1.84
1768	-6.0	4.3	7.9	11.1	7.0	7.1	8.2	8.2	6.2	7.8	0.3	-6.0	4.65	7.38	1.04	-0.09
1769	-5.0	-5.0	3.6	5.7	15.2	2.6	5.7	9.1	4.8	0.5	-4.5	-7.0	2.12	6.29	-1.08	-3.11
1770	-3.5	-0.6	4.8	4.9	6.6	8.2	5.0	6.8	7.2	4.6	-2.4	-6.5	2.91	6.38	-0.99	-1.59
1771	-4.5	6.2	7.4	8.3	10.5	8.7	6.8	7.2	6.7	7.9	-1.2	-3.0	5.07	7.96	1.40	0.63
1772	-2.4	1.4	9.9	8.2	7.3	9.8	5.8	5.0	4.6	2.0	-4.5	-7.5	3.28	5.73	2.14	1.46
1773	-2.4	-0.9	7.0	6.5	13.2	9.0	7.6	6.3	8.8	5.9	-4.5	-8.5	3.99	8.46	-0.31	-0.96
1774	-3.5	0.7	4.2	2.1	8.7	12.2	5.8	5.2	7.0	6.2	-5.5	-8.5	2.86	7.48	-1.59	-1.79
1775	-6.5	-1.2	3.9	11.4	12.6	11.0	7.6	5.2	7.7	2.0	-4.5	-6.5	3.54	7.65	-1.07	-3.08
1776	-0.3	1.7	5.2	13.4	10.5	10.6	8.1	6.2	6.8	3.5	-4.5	-4.5	4.49	7.58	1.48	0.00
1777	-3.0	-4.5	7.2	7.6	11.1	8.9	9.1	8.5	6.9	4.5	-7.0	-7.0	3.86	8.13	-0.72	-1.21
1778	-6.0	-1.8	6.4	9.7	9.1	10.7	7.3	6.2	2.8	5.6	-2.7	-2.7	3.39	6.94	-0.24	-2.11
1779	-2.1	3.1	5.6	7.7	8.9	6.2	6.4	6.6	6.7	4.6	-6.5	0.6	3.93	6.54	0.84	0.96
1780	-1.2	2.7	5.3	5.9	5.7	8.0	6.0	5.3	7.4	4.2	-7.0	-3.5	3.66	6.08	1.06	1.86
1781	-5.5	-0.6	1.9	8.3	7.6	8.2	8.0	4.9	7.2	5.7	-1.8	-5.5	3.24	6.93	-0.21	-1.92
1782	-4.5	-0.3	9.3	1.9	10.1	2.5	4.3	1.2	5.1	2.3	-1.2	-4.5	1.91	4.25	-0.05	-0.25
1783	-0.6	1.5	6.4	3.4	10.4	9.2	5.9	5.5	8.0	2.1	-4.5	-2.4	3.36	6.83	0.29	0.71
1784	-0.6	-0.9	2.6	0.9	8.5	7.0	8.3	4.2	9.2	-0.3	-9.0	-6.5	1.99	6.14	-1.56	-0.32
1785	-3.0	-1.5	7.7	8.7	16.2	7.7	9.3	4.6	8.7	2.2	-8.5	-8.5	3.74	8.09	-0.53	-0.84
1786	0.0	-1.5	3.3	11.0	5.1	11.1	3.9	6.7	13.3	9.8	-7.0	-8.5	6.25	8.28	-0.46	-1.68
Avg.	-3.53	-0.11	5.61	7.00	9.57	8.28	6.42	5.48	7.00	4.28	-4.73	-5.58	3.38	6.84	-0.15	-0.86
St. Dev.	1.96	2.52	2.39	3.23	2.66	2.34	1.84	1.89	1.94	2.46	2.34	2.68	1.01	1.08	1.10	1.41

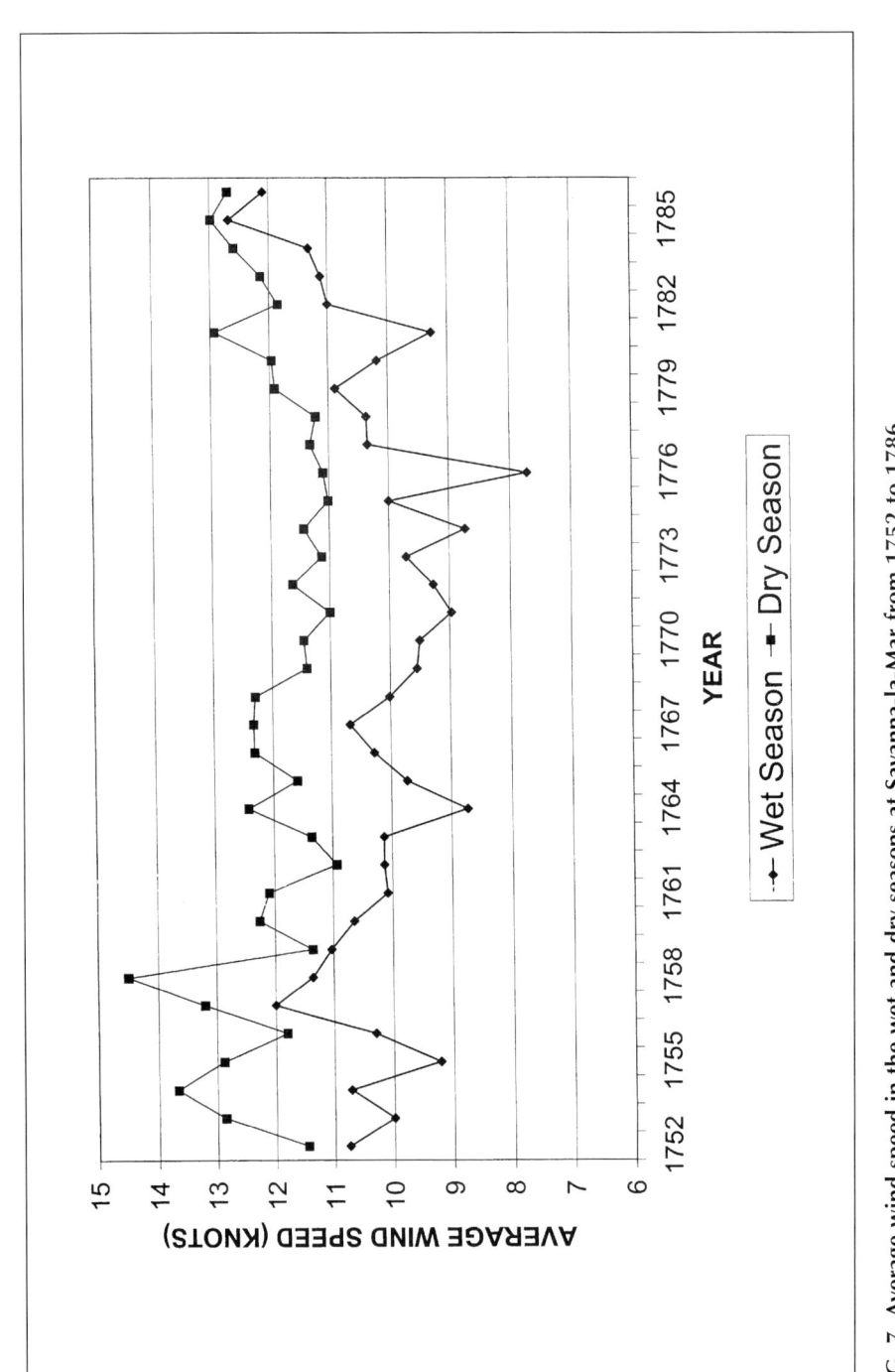

FIG. 7 Average wind speed in the wet and dry seasons at Savanna-la-Mar from 1752 to 1786.

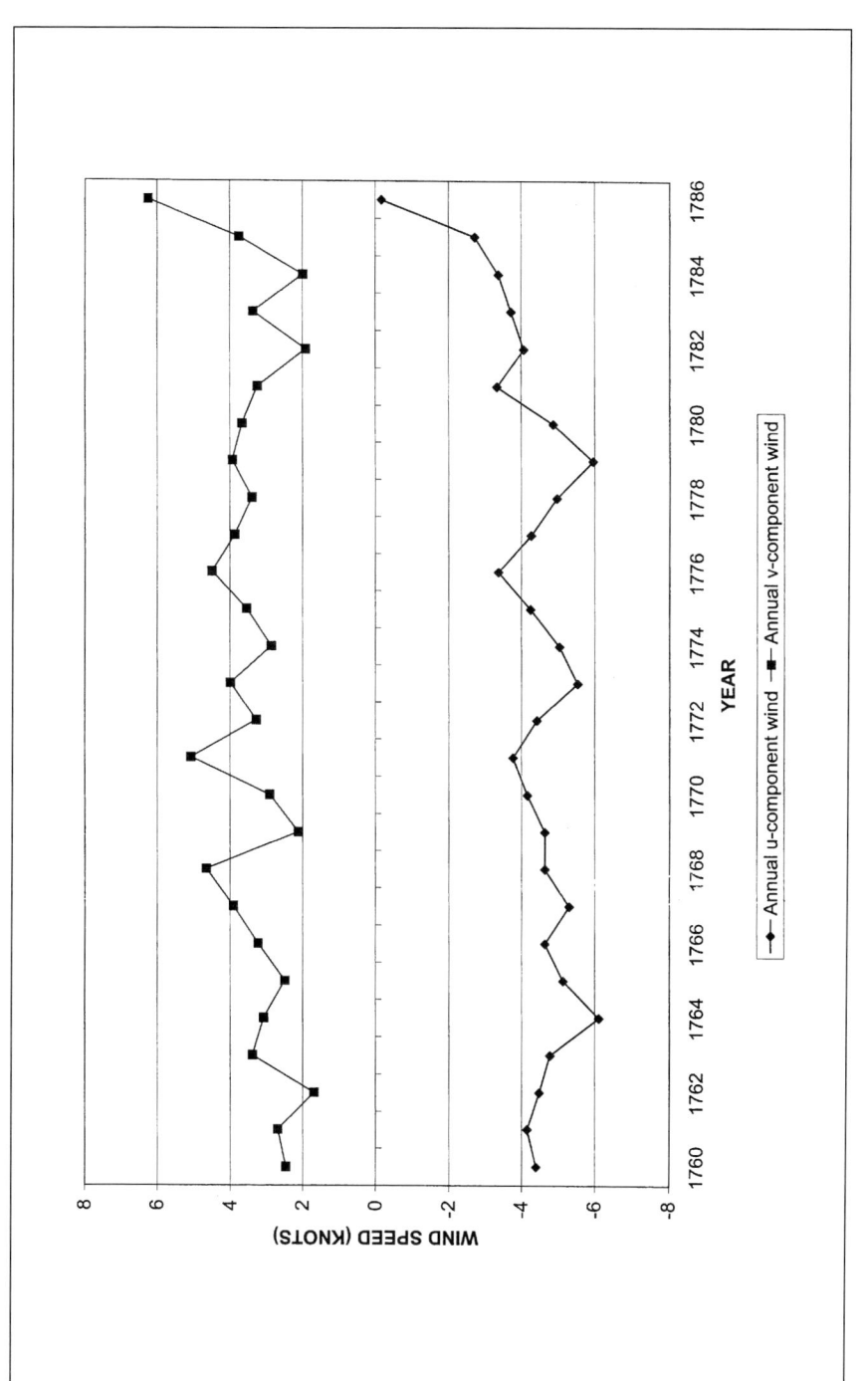

FIG. 8 Average annual u- and v-component winds at Savanna-la-Mar from 1760 to 1786.

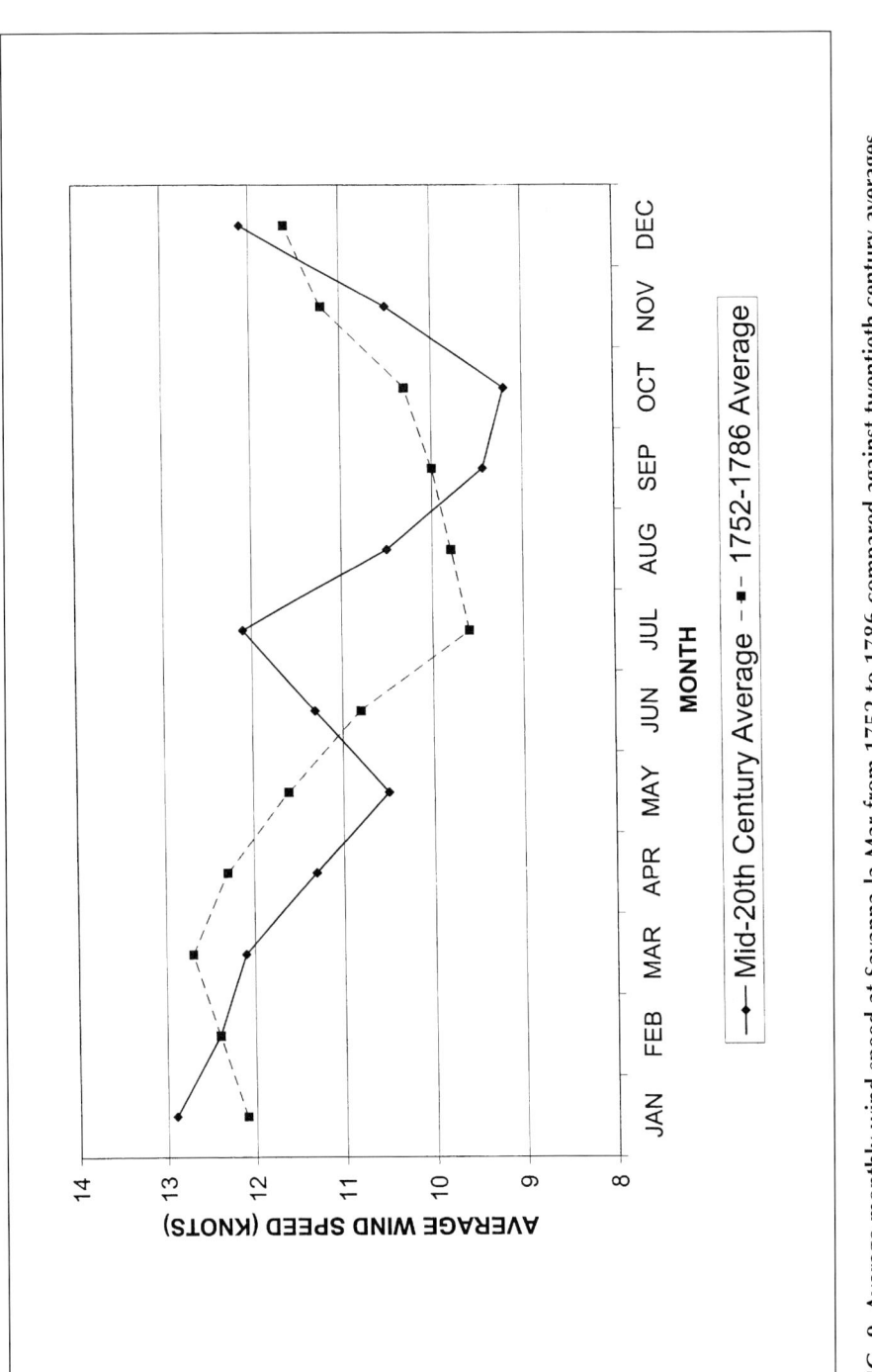

FIG. 9 Average monthly wind speed at Savanna-la-Mar from 1752 to 1786 compared against twentieth century averages.

INTERPRETATION OF RESULTS

The results indicate that the middle and late eighteenth century was cooler and moister than in the twentieth century. The cooler temperatures and enhanced dry season rainfall indicates that cold fronts penetrated more frequently and further to the south in the eighteenth century. Cold fronts account for much of the dry season rainfall along the major islands of the Greater Antilles. During the wet season, the slightly lower wet season rainfall is in line with lower temperatures and reduction in the amount of water vapor that can be held in the atmosphere at lower temperatures.

It is unfortunate that the wind speed data cannot be reliably compared with modern data. Higher wind speeds were observed in the 1750s and 1780s, and the number of thunder days was reduced in both periods. These results are consistent because higher surface wind speeds are usually correlated with a more stable atmosphere that is less favorable to thunderstorm formation.

Most surprising is the evidence (see Fig. 9) that there is no July maximum in wind speed in the eighteenth century. The maximum is not shifted to another month in summer; rather it is not present at all. The July maximum in wind speed is associated with the seasonal strengthening and expansion of the North Atlantic high pressure centered near the Azores. The absence of this midsummer increase in wind speed suggests that the North Atlantic high was weaker than in the mid-twentieth century. This would produce weaker Trade Winds and relatively higher temperatures. It also suggests that atmospheric pressure was lower, and therefore vertical shear was decreased and the Trade Wind inversion weakened, hence a moister atmosphere conducive for the development of organized convection and possibly tropical cyclones (Knaff 1997). As will be shown in Chapter 5, tropical cyclone numbers in the Jamaican region from 1777–1790 were higher than at any time from 1750–2002.

The available temperature record covers only the period when wind speeds were generally below the 1752–1786 average. The absolute minimum in wind speed did coincide with the period of highest air temperature from 1770–1773. This in turn coincided with the minimum in rainfall at Savanna-la-Mar as indicated both in the amount (1760–1786) and number of rain days (1752–1786). While even lower temperatures might be expected in the windier 1750s and 1780s, the documented effects of warm El Niño/ Southern Oscillation (ENSO) events can also influence temperature in the Jamaica area. However, a minimum in estimated sea surface temperature in Puerto Rico in 1780–1785 (Winter et al. 2000) is not refuted by the Jamaican data.

The warmer temperatures in 1770–1773 (and perhaps also in 1769) coincide with the worst extended drought in Savanna-la-Mar and the worst drought island-wide up to that time (Long 1774). There is no documented warm ENSO from South American records (Ortleib 2000), but severe drought and famine occurred in India in 1769–1770 and may indicate a warm ENSO event with global effects (Whetton et al. 1996). The Jamaican wind speed data was near its minimum at this time, and lower wind speeds favor increased sea surface and air temperature due to reduced evaporation and ocean surface mixing. However, as other adjacent years had similarly low wind speeds, it seems likely sea surface temperatures were higher than usual in the tropical eastern Pacific Ocean.

In the modern record, during the months of December through March, there is a correlation between anomalous wind flow over the Caribbean and the Gulf of Mexico region and ENSO (Enfield and Mayer 1997).[22] However, Harrison and Larkin (1998) have shown that this region does not consistently respond in different warm ENSO events. Instead, more frequent low pressure centers, or frequent troughs of low pressure (cold fronts) will, on average, produce lower pressure in the region and weaken the winter Trade Wind regime in Jamaica.

Examination of the u and v-component wind data for these months reveals enhanced southwesterly flow in the winters of 1767–1768 and 1770–1771 while there was a distinct northeasterly anomaly in the winter of 1768–1769 (see Table 20). Both 1768–1769 and 1770–1771 were the two most distinct events, indicating more frequent high pressure affecting Jamaica in the winter of 1768–1769 and weaker high pressure to the north of Jamaica in the winter of 1770–1771. However, there is no incontrovertible evidence of a cold or warm ENSO event based only on the wind data.

The temperature data suggest that the sequence of events may actually have been a warming of eastern Pacific sea surface temperature (SST) in early 1769 that persisted at least into 1771. Severe drought in Jamaica, Mexico (*South Carolina Gazette*, 9 August 1770) and India (Whetton et al. 1996) during all or part of 1768–1770 is suggestive of a global-scale phenomenon. Continued warmth in Jamaica through 1773 may indicate that eastern Pacific

---

[22] The modern data is derived from the NCAR/NCEP Re-Analysis Project (Kalnay et al. 1996) and available from the NOAA Climate Diagnostics Center web site: *http://www.cdc.noaa.gov*.

SST remained warmer than average even if there were no signifi-cant El Niño warm events along the coast of Ecuador and Peru. An analogy may be that of extended warmth in the eastern Pacific in 1990–1995 (Allan and D'Arrigo 1999), during which only one winter featured a classic warm event along the coast of South America (Harrison and Larkin 1998).

Other winters exhibiting enhanced southerly flow in Jamaica are 1759–1760, 1775–1776, 1778–1779, 1779–1780, 1781–1782, 1782–1783, 1783–1784 and 1785–1786 (see Table 20, last col-umn). Other winters with enhanced northerly flow in Jamaica are 1762–1763, 1763–1764, and 1774–1775.[23] The only firmly docu-mented warm ENSO event from 1760 to 1786 is the winter of 1784–1785. There may have also been an event in either 1759–1760 or 1760–1761 (Ortleib 2000). However, the dating of El Niño events to the winters of 1760–1761, 1767–1768 and 1778–1779 by Garcia Rodriguez fall in winters with enhanced south or southwesterly flow in Jamaica or at least no anomalous northeasterly flow. There is independent confirmation from other records included in the Ortleib compilation, which in light of the new Jamaican data strengthens the case for warm ENSO events in 1760–1761, and in 1768. This also suggests that Garcia Rodriguez may be a reliable reporter of historical El Niño events.[24]

The extended period of enhanced southerly winter winds in Jamaica from the late 1770s to 1786 indicates more frequent pushes of cold air into the southern portions of North America (i.e., the southeast United States). The winters of 1779–1780 and 1783–1784 are documented as two of the coldest winters ever to affect the eastern half of the United States. The winter of 1784–1785 was also extremely cold (Ludlum 1966). But the win-ters of 1780–1781, 1781–1782 and 1782–1783 were all milder than average (Landsberg et al. 1968). Compared to the long-term average, only the winters of 1780–1781 and 1782–1783 are notably mild. This would indicate that more frequent pushes of cold air into Jamaica from the north were more common at this time.

---

[23] The winter of 1774–1775 was the warmest winter of the 1770s in Virginia. The combination of enhanced northeasterly winds and below-average temperatures in Jamaica, and above average temperature and assumed enhanced southwest-erly flow, suggests a northwestward-displaced North Atlantic high pressure cen-ter dominating the Gulf of Mexico and southeast U.S. region. My thanks to Dennis Blanton, The Center for Archaeological Research, College of William and Mary, Williamsburg, Virginia, for providing the temperature data.

[24]Ortleib (2000) indicates that he was unable to locate this source for analysis of its reliability.

This is supported by (very rare) comments made by Thistlewood in his weather journal in March 1786 and is in evidence in the anomalous meridional wind components observed in most of 1786. On 4 March, he wrote: "Mod[era]te Breezes NW & at times Cloudy. A very uncommon Weather & Wind." Two days earlier he had commented on moderate breezes from the southwest that blew all night "which [I] never remembered in this Island before."[25]

There is also a positive correlation between May–September scalar wind speed in the Jamaican region and SST in the equatorial eastern Pacific Ocean. This correlation is highest south of Jamaica in the zone of maximum average wind speeds. The increase in eastern Pacific SST is frequently linked with warm ENSO conditions. Because sea-level pressure near the equator undergoes small absolute changes, the changes in wind speed in Jamaica are evidence for either higher pressure in the Atlantic high-pressure cell or abnormally low pressure (almost always associated with warm ENSO conditions; see Harrison and Larkin 1998) in the tropical Eastern Pacific, or both. But Harrison and Larkin (1998) did not find a significant statistical correlation in Caribbean Sea summer season Trade Winds and eastern Pacific SST in the 1948–1993 period.

Each May–September period from 1782 through 1786 showed increased scalar wind speeds. However, the increase may be related not only to ENSO but also to a southward displaced and/or strengthened North Atlantic high-pressure center. U.K. Navy ships' logbooks for the hurricane season of 1782 in the Leeward Islands indicate no hurricanes or tropical storms that season. This suggests that 1782 was a season of increased trade wind strength and a possible warm ENSO event in the tropical Pacific. The situation is less clear-cut in subsequent years.

Historical records of the frequency of westerly winds over the British Isles indicate that the lowest number of days with westerly winds were in the 1780s (Lamb 1977). Meridional circulation was enhanced in the 1780s, and the North Atlantic high was displaced to the south and west of the United Kingdom. Evidence for this also comes from the Netherlands in 1785, where the normally moist westerly wind regime was so infrequent, that the foundations of the sea walls were exposed for the first time since their construction in the previous century (*Pennsylvania Gazette*, 7 September 1785). The next time that the number of westerly wind days in the United Kingdom reached such low numbers was in the

[25] Thistlewood, Monson Collection 39/72.

early and mid-1970s. (Lamb 1977). Weakened westerly winds over the United Kingdom are positively correlated with the North Atlantic Oscillation (NAO) Index. The NAO Index is a measure of the strength of the prevailing westerly wind flow over the North Atlantic, which is determined by the pressure gradient between the Iceland Low and the Azores High (Hurrell 1995). Changes in the strength of the Azores High in turn affect the strength of the Atlantic Trade Winds that blow along its southern fringe. However, it should be noted that the NAO Index is most highly correlated in the winter months and least in the summer months.

If a similar situation prevailed in the 1780s as it did in the 1970s, then we would expect lower temperatures and stronger Trade Winds in the Caribbean region. There is evidence for cooler temperatures in the 1780–1785 period from Puerto Rican corals (Winter et al. 2000). There is only one confirmed warm ENSO event in the 1780s (1784–1785) (Ortleib 2000). The weight of the available evidence would implicate the southward displacement of the North Atlantic high as the greater contributor to increased Trade Wind strength at Jamaica in the wet season of the 1780s than increased eastern Pacific sea surface temperatures.[26]

Finally, it should be noted that the NAO appears connected with temporal variability in the tropical Atlantic that stands out at the 10–20 years' time scale (Hansen and Bezdek 1996; Chang et al. 1997). The changing strength of the Trade Winds at Savanna-la-Mar from a maximum in the 1750s to a minimum in the 1770s followed by renewed strengthening in the 1780s suggests a similar variability.

CARIBBEAN CLIMATE IN THE LATE EIGHTEENTH CENTURY

The climate of Jamaica in Thistlewood's time, compared to 200 years later, was cooler and moister. Of course, it was still a hot, moist tropical location, and an Englishman in either time would be hard put not to make a distinction between the climate of each

---

[26] However, if there were other warm ENSO events at this time then 1782–1783 is a very good candidate. Although no direct evidence is available from the coast of Peru and Ecuador, there is evidence of drought and monsoon failures in India at this time (Whetton et al. 1996), a severe wet season drought in Jamaica, and drought-induced fires that raged for five weeks on the island of Trinidad. During the 1820–1821 warm ENSO event, there was a three-week period of fires that were considered the most severe on the island of Trinidad since 1782 (*London Times*, 22 October 1821). Jamaica also suffered severe drought in the 1820–1821 warm ENSO event (*London Times*, 24 October 1821).

location. However, good quality weather records are required to make meaningful and accurate comparisons of climate over the centuries.

Several factors explain the change in climate. First, there has been a gradual warming of the Earth's climate for the better part of 150 years or more in most regions of the world. Second, the circulation patterns of the North Atlantic basin were weaker in the eighteenth century than in the twentieth century. The main storm tracks were shifted farther to the south, allowing cold air to penetrate farther south. At the same time, the intertropical convergence zone in the Americas and West Africa, on average, seems to have shifted farther north as the Azores High weakened. This pattern brings more rainfall to the Sahel region of West Africa, and by limiting the westward expansion of the Azores High in the wet season, allowed for moister, less stable conditions to prevail in Jamaica.

At the same time, the tropical eastern Pacific Ocean may have been much colder than in the twentieth century. Relative to the tropical Atlantic, where weaker Trade Winds are implicated, colder temperatures in the Pacific may indicate a stronger South Pacific High and stronger Trade Winds favor lower air and sea surface temperatures. A pattern of weak Atlantic Trade Winds and relatively higher temperatures and a cold equatorial Pacific Ocean favors enhanced rainfall in Central America and the Caribbean region today. The weaker Azores High allows for an enhanced southwest monsoon flow across Central America. An unusually warm tropical Pacific (El Niño conditions) and an unusually cold tropical Atlantic (strong Azores High) will bring unusually dry weather to Central America and the Caribbean (Enfield and Mayer 1997; Enfield and Mestas-Nuñez 2000).

Weaker westerly winds in the North Atlantic and weaker Trade Winds in the Tropical Atlantic in the eighteenth century, combined with a colder equatorial Pacific Ocean, is a scenario that would allow the possibility of enhanced tropical cyclone activity in the Tropical Atlantic. Chapter 5 presents the evidence in support of this scenario.

chapter 5

HURRICANES AND TROPICAL STORMS

INTRODUCTION

Perhaps the most important new information to be derived from the Thistlewood weather journals is for tropical cyclones. Current compilations of Atlantic basin tropical storms and hurricanes are known to be incomplete. Tannehill ([1938] 1952)[27] collected all published accounts from the major compilations of historical tropical cyclones in the North Atlantic from 1492. He omitted storms from the winter months, which could not be tropical in origin, but otherwise did not attempt to assess the accuracy of the information. Ludlum (1963) studied American newspapers, local histories, and other published accounts to add to our knowledge of outstanding major hurricanes along the Atlantic and Gulf coasts of the United States. Millás (1968) reexamined the compilations of previous Caribbean storms for the period 1492–1800 and added new information from Spanish sources. Millás performed a critical assessment of each case study and was able to dismiss many previously accepted storms, corrected inaccuracies in other accounts he subsequently accepted, and added new storms.

The earliest compilers do not always provide the source for their information, but much of the earliest data come from letters, government documents, annual registers, and history books, and in later years, from newspaper and magazine accounts. The first systematic scientific investigations of hurricanes date from the early nineteenth century, with William Redfield and William Reid leading the way in the collection of newspaper reports and ship logbook weather data. Their data allowed the depiction of tropical cyclones in map form. Theoreticians engaged with them and other scientists to use this new information to further develop competing theories

[27] I use the 8th edition, published in 1952.

about the nature and origin of hurricanes and other atmospheric storms (Fleming 1990).

Today, tropical meteorologists have an impressive array of tools for monitoring and tracking hurricanes and tropical storms. Satellites provide continual and complete survey of the world's oceans, ensuring that even the weakest storms no longer go undetected. Weather stations on land and ships at sea provide surface weather data. Upper air wind data (critical for forecasting tropical cyclone movement and intensity) are gathered by ground-launched weather balloons. Satellites provide data about the upper air.

In the North Atlantic, continuous and complete satellite coverage dates from 1966, but the record back to 1944 is also abundant. Regular aircraft reconnaissance of tropical cyclones began in 1944 (Landsea et al. 1999), and the total number of surface weather reports from land and sea was greater in the late 1940s to mid 1960s than is currently available. Satellite data now help to fill in gaps in the ground-based observing network. During the period 1944–2001, a total of 586 tropical cyclones occurred (i.e., storms with sustained wind speeds of at least 39 knots near the center of the storm). Of these 586 storms, 345 intensified into hurricanes (sustained wind speeds of at least 64 knots).[28] This represents an average of 10.1 tropical cyclones each hurricane season (June through November), of which, on average, 5.9 develop into hurricanes.

Jamaica lies in the southwest region of the North Atlantic. This area of the tropical Atlantic has an early season (late May through June) peak in tropical storm development. Activity diminishes in July and then peaks again in the main hurricane months of August through October. The month of October is particularly prone to tropical storm development in the western Caribbean. Unlike the June maximum of activity, storms in October are much more likely to develop into major hurricanes. The highest risk of major hurricanes in Jamaica is from the beginning of August to mid-November.

North Atlantic tropical cyclones vary in number and intensity from one season to the next. The El Niño/Southern Oscillation (ENSO) phenomenon in the Pacific Ocean is known to have a marked effect on the number and intensity of hurricanes in the Atlantic Ocean. Changes in stratospheric winds (Quasi-Biennial Oscillation) are also implicated in inter-annual fluctuations of

---

[28] All modern hurricane data are available from the National Hurricane Center Web site at www.nhc.noaa.gov (Jarvinen et al. 1984).

Atlantic hurricanes (Gray 1984; Bove et al. 1998). Changes in atmospheric pressure and sea surface temperature within the Atlantic basin itself, and of rainfall in the western Sahara of Africa, are also implicated (Gray 1984; Elsner and Kara 1999). Recent research also reveals that there are long-term variations, which can persist for several decades, in the total number and intensity of storms (Gray 1990; Landsea et al. 1992, 1999). From the mid-1960s to the mid-1990s, hurricane numbers and intensity were much below that of the period from the early 1930s to the early 1960s (Goldenberg and Shapiro 1996; Landsea et al. 1999; Elsner and Kara 1999). The most complete records back to 1886 suggest other periods of relative low and high activity before the early 1930s. A longer-term perspective on North Atlantic tropical cyclone variability was the subject of a recent workshop attended by specialists in historical hurricane reconstruction.[29]

Based on the best climatology (1944–2001) a representative 35-year period would have about 354 named tropical cyclones, of which about 207 would develop into hurricanes throughout the entire North Atlantic basin. The centers of 15 tropical cyclones have come within 100 miles of Savanna-la-Mar in the years 1944–2002. This number is considerably below that of 40 storms in the years 1871–1943. Twelve of these hurricanes passed close enough to Savanna-la-Mar to produce hurricane force winds. Nine occurred from 1871–1943 and only three from 1944–2002. This is all the more remarkable when we consider that the pre-1944 period is likely to underestimate the total number of North Atlantic tropical cyclones, and certainly the maximum intensity of those observed (Landsea et al. 1999).

The Thistlewood record allows a continuous record from a single site to be reconstructed for the years 1750–1786. For July 1750–July 1751, the record is for Black River, about 20 miles southeast of Savanna-la-Mar. After this date, Thistlewood lived near Savanna-la-Mar. His records provide partial coverage for the 1750 and 1751 hurricane seasons, and a continuous record for the 1752–1786 seasons. This provides a homogeneous record of wind force that can be used to estimate the intensity of each tropical cyclone as felt at Savanna-la-Mar. The maximum intensity of each tropical cyclone cannot be derived from a single point location, but it can give information on the risk at a single point and serve as a point of comparison with the risk in twentieth century records.

[29] Based on discussion groups and recommendations of the Workshop on Atlantic Basin Paleohurricane Reconstruction from High Resolution Records, University of South Carolina, Columbia, SC, 25–27 March 2001.

METHODS

Because Thistlewood did not have a barometer, evidence for tropical storm force (and higher) winds comes from his observations and descriptions of wind direction and force, general weather, state of the sea, and deviations of the weather from its average state. These data are informed by inferences about the movement and climatology of tropical cyclones in the twentieth century (Neumann et al. 1999).

Wind directions between southwest and northwest during the months of May through December were extracted from the database. Because these wind directions are unusual at any time in the year they offer evidence of a closed center of low pressure.

Winds blowing more than 45 degrees from the resultant direction for day (170 degrees) and night (45 degrees) that were sustained throughout an entire day or night period were also flagged because they provide evidence that the prevailing wind field is abnormal. For example, during the hurricane season, the persistence of a northerly wind during the day (when a southerly breeze should be blowing) is evidence for a tropical depression, tropical storm, or hurricane being located to the east.

All winds described as "strong gales" or of greater force were flagged. In the months of November and December, the southward movement of cold polar air masses can cause gales. An assessment of the actual weather situation can often be determined from the other information in Thistlewood's weather observations. In other instances, independent data were used to make a judgment on the nature of the weather. Sustained winds of "strong gale" strength are not considered to be reliable indicators of a tropical storm in the area, but are strong enough to be flagged for consideration. Winds higher than "strong gales" are considered to indicate wind speeds of tropical storm or greater force.

The roaring of the sea was noted at times by Thistlewood, and inevitably (during the hurricane season) was associated with hurricanes that either affected the island or passed by at some distance.

General weather descriptions of wind squalls,[30] violent rains, sustained rains, and the like provide information that assist in identifying tropical cyclones. Particularly important is the distinguishing

---

[30] Squalls are sudden violent rushes of wind with a wind speed of 18 miles per hour or higher. The squalls may only last a few minutes and then be followed by intervals of little or no wind. In tropical cyclones, stronger squalls are embedded within the rain bands that surround the tropical cyclone center. Winds are frequently lighter outside of the rain bands.

of severe local thunderstorms of short duration that can produce violent winds. Knowledge of the local climatology, and the diurnal patterns of the weather, is of great importance.

RESULTS

Frequency of Tropical Storm and Hurricane Force Winds at Savanna-la-Mar

Table 21 lists the twelve tropical cyclones that produced sustained winds of tropical storm or hurricane force and/or tropical storm or hurricane-force squalls in the Thistlewood record from 1750–1786. The observed prevailing wind direction(s), maximum sustained wind force, and highest gusts are given, and the storms are ranked in their relative magnitude from strongest (top) to weakest (bottom). Other storms passed far enough away from Savanna-la-Mar to produce only lighter and briefer episodes of wind squalls. Only five storms produced sustained winds of hurricane force and/or hurricane-force squalls (wind force 10 or higher in Table 21). One year, 1781, had one tropical cyclone that produced sustained tropical storm-force winds, and two others that produced tropical storm-force gusts. Eight of the 12 storms were in the nine-year period 1778–1786 and seven were in the seven-year period 1780–1786.

Table 22 provides an estimate of the number of tropical storms that passed within 100 miles of Savanna-la-Mar for the years 1871–2001. The use of a 100-mile radius to estimate hurricane frequency and intensity probabilities for a given location as defined by T. Kimberlain[31] is best applicable to the wind force estimates in the Thistlewood record.

The date of nearest approach to Savanna-la-Mar and the estimated maximum wind speed near the storm center are included. The final column is not in the North Atlantic hurricane database, or HURDAT (Landsea et al. 1998) and provides a subjective estimate of the most likely maximum sustained winds at Savanna-la-Mar. These are given as either tropical storm force (T) or hurricane force (H). If sustained tropical storm-force winds are considered not to have occurred at Savanna-la-Mar, then "NONE" is entered. The entry "Gusts" indicates that peak wind gusts of tropical storm force were likely, but no sustained winds of this force probably occurred. Fifty-five tropical cyclones passed within 100 miles of Savanna-la-Mar from 1871–2002. Forty-two of them are

[31] Available at *www.aoml.noaa.gov/hrd/tcfaq/tcfaqG.html#G12.*

estimated to have produced tropical storm-force or stronger winds at Savanna-la-Mar.

There were definitely three, and very likely five, instances of sustained hurricane force winds at Savanna-la-Mar from 1750–1786. Two other storms produced gusts (force 9) approaching hurricane force. Three (or possibly five) hurricanes in 37 years give an average frequency of one hurricane every 12.3 (7.4) years. The total number of days with tropical storm-force or greater winds at Savanna-la-Mar in 1751–1786 was 12 (one every 3.0 years). This is nearly identical to the same estimated value for 1871–2002 of one every 3.1 years (42 in 132 years).

From 1871–2002, there were far more frequent tropical cyclones before 1953 than afterward. Only eight tropical cyclones passed within 100 miles of Savanna-la-Mar from 1954–2002. This is a rate of only one storm per 6.1 years. From 1903–1953, the rate was once every 1.5 years. The period 1944–2002 is considered to be the time of most reliable tropical cyclone monitoring. The average in this period is once every 4.0 years.

Considering only hurricanes passing within 100 miles of Savanna-la-Mar, twelve hurricanes are estimated to have produced hurricane-force winds at Savanna-la-Mar from 1871–2002. Nine were before 1944 and three occurred since 1944. For the

TABLE 21

Dates of sustained tropical storms and hurricane force winds at Savanna-la-Mar from 1751 to 1786. DDD is the wind direction in degrees where 360 indicates north. FF is the wind force category from Thistlewood's personal scale. FQ is the frequency of squalls. A '0' indicates no descriptor of squalls, a '2' indicates 'some squalls', a '3' indicates frequent squalls and a '4' indicates many squalls. An 'X' in the final column indicates that the roaring of the sea was recorded. In two instances, no wind direction was given during the time of the strongest winds.

Year	Month	Day	DDD	FF	FQ	SQ	SEA
1780	OCT	3	180–200	12	4	>12	X
1751	SEP	22	110	10	4	12	X
1781	AUG	2	020–180	10	4	11	
1764	OCT	3	135	8	3	10	
1786	OCT	20	360–135	8	3	10	
1784	JUL	31	020–160	6	3	9	
1778	SEP	16	070–045	6	0	9	
1766	AUG	16	180–110	6	0	8	X
1781	SEP	6	170	6	3	7	
1785	AUG	28	135	6	3	7	
1756	SEP	17		6	2	7	
1781	OCT	8		6	0	7	

TABLE 22

Dates of tropical cyclones that passed within 100 miles of Savanna-la-Mar from 1871 through 2002. The maximum sustained wind speed and the storm type (T for tropical storm and H for hurricane) of each tropical cyclone is followed by the subjectively estimated maximum sustained wind at Savanna-la-Mar (between 40 and 73 miles per hour for a tropical storm and greater than 73 miles per hour for a hurricane). "None" indicates that no sustained tropical storm-force or greater winds were probably felt at Savanna-la-Mar.

Year	Date(s)	Storm Center Max Speed	Type	Storm Type Storm at Savanna-la-Mar	Year	Date(s)	Storm Center Max Speed	Type	Storm Type Storm at Savanna-la-Mar
1874	7–Oct	80	H	T	1931	14–Sep	55	T	T
1879	13–Oct	50	T	T	1932	30–Sep	40	T	T
1880	6–7 Aug	80	H	T	1933	16–17 Jul	45	T	None
1880	18–19 Aug	70	T	T	1933	17–Aug	45	T	None
1886	28–Jun	35	T	None	1933	21–Sep	65	T	T
1886	21–Aug	85	H	H	1933	30–Oct	85	H	H
1886	17–18 Sep	85	H	T	1934	20–Oct	40	T	T
1887	16–Sep	85	H	T	1935	28–Sep	100	H	H
1889	16–Sep	85	H	T	1935	20–Oct	60	T	T
1893	6–Nov	35	T	None	1938	13–Aug	75	H	T
1895	26–Aug	85	H	T	1939	3–Nov	55	T	T
1896	27–Sep	105	H	H	1942	20–Sep	45	T	T
1898	30–31 Oct	50	T	T	1944	21–Aug	105–70	H	H
1903	12–Aug	95–100	H	H	1944	13–15 Oct	70	T	T
1904	14–Jun	40	T	None	1947	21–Sep	35	T	T
1904	14–Oct	35	T	None	1948	20–Sep	50	T	None
1906	8–Nov	35	T	None	1950	15–16 Oct	55	T	T
1909	16–17 Jul	45	T	T	1951	17–18 Aug	75	H	T
1909	23–24 Aug	90	H	H	1953	24–Sep	40	T	T
1910	10–Sep	70	T	T	1974	1–Sep	75–80	H	T
1912	15–19 Nov	130	H	H	1974	17–Sep	35	T	None
1915	12–13 Aug	95–100	H	H	1980	6–Aug	115	H	H
1916	15–16 Aug	95	H	H	1988	13–Sep	110–115	H	H
1916	30–31 Aug	80	H	T	1994	12–13 Nov	35	T	T
1917	24–Sep	95	H	T	2001	8–Oct	75	H	T
1924	7–8 Nov	40	T	T	2002	19–Sep	45	T	None
1927	19–Oct	40	T	None	2002	30–Sep	50	T	None
1928	3–Sep	35	T	None					

years 1871–1943, this is an average of one hurricane every 6.1 years. For 1944–2002, the average is one hurricane every 19.3 years.

From 1750–1786, five different days brought hurricane-force winds to Savanna-la-Mar. This gives an average of one hurricane every 7.4 years. These figures fall within the observed range in 1871–2001, but are much closer to the average of 1871–1943 than 1944–2001.

Thistlewood did not suffer a severe hurricane during the years 1752–1779, but he experienced significant property damage in 1778, 1780, 1781, 1784, 1785, and 1786. The hurricanes of 1780, 1781, and 1786 were the worst, with the 1780 hurricane destroying virtually all his property. The last five hurricanes came in a seven-year period, and such a quick succession of storms in the island of Jamaica is only documented in 1812–1818 (1812, two in 1813, 1815, 1818) (*Jamaica Royal Gazette*, various issues). Three hurricanes produced hurricane-force winds at Savanna-la-Mar in a seven-year period (1780–1786; 1812–1818, and 1912–1917). From 1909–1917, six hurricanes passed within 100 miles of Savanna-la-Mar, but the 30–31 August 1916 and 23 September 1917 storms are estimated to have not brought hurricane force winds to Savanna-la-Mar. The one in 1912 was comparable to the great 1780 hurricane. From 1780–1786, there were five hurricanes that are estimated to have passed within 100 miles of Savanna-la-Mar. The highest in the years 1871–2001 is six, in the nine-year period 1909–1916 (of which five occurred in a six-year period, 1912–1917).

THE IMPACT OF TROPICAL CYCLONES ON EIGHTEENTH-CENTURY JAMAICA

What impact did tropical cyclone variability have on Thistlewood and his neighbors? Thistlewood was fortunate in his timing. He witnessed a major hurricane in his second year on the island and learned the potential risk and actual costs of such storms. By the time he was able to purchase his own estate, sixteen years had passed without major damage to property in the area. He was again fortunate to build his estate during a period of unusually low tropical cyclone activity, a period including the worst drought known in Jamaica since the English seizure of Jamaica in 1655 (Long 1774). In this sense, the sugar planters in Jamaica were also fortunate that there was an extended period of low hurricane landfall frequency at just the time the sugar industry took off.

What Thistlewood built was almost nearly all lost in the great hurricane of October 3, 1780 and rebuilding efforts were hampered by subsequent storms until his death. However, Thistlewood was not a big sugar planter, but a man of more modest means. For the large planters, sugar remained a profitable business, and great amounts were produced each year from 1788 to 1790, indicating the adaptability of Jamaican planters to the worst weather that could be thrown at them. Clearly, enough money could be made at this time that even frequent hurricanes could not long suppress profit. The later peak in activity in the early nineteenth century may not have been handled as well because the international market conditions for all agricultural products were likely reshaped by two decades of war and the decline of the slave trade in the English-speaking islands. The weather was probably not a determining factor for the sugar industry as a whole, but planter costs to repair damage would not be desirable at any time.

Thistlewood's daily weather record for 1750–1786 may, in future, be used with some profit to examine the effect of weather conditions on sugar and other crops for each year. At various critical points in the growing season, the timing, frequency, and amount of rainfall will retard or favor growth and the final yield of the crop. Such effects can then be accounted for and removed from long-term trends in planted acreage, war and other factors known to influence planter decisions. The summarized statistics in Chapter 4 provide a useful start toward such a study, as does the list of tropical storms and hurricanes known to affect Savanna-la-Mar, shown in Appendix 4.

LONG-TERM TROPICAL CYCLONE VARIABILITY IN JAMAICA

Tropical Cyclones Passing Within 200 miles of Savanna-la-Mar

Table 23 is a list of all tropical cyclones centered within 200 miles of Savanna-la-Mar for the period 1750–2002. The data for 1871–2002 are exclusively from the HURDAT series. Data for 1851–1870 includes new data gathered by the author and merged with available (but incomplete) HURDAT data for these years. For 1750–1850, the data are from new sources gathered by the author, most of it never previously used. The data sources include the Thistlewood record, newspapers, and ships' logbooks. A complete compilation of these storms, along with a critical analysis of previously accepted tropical cyclones, will be published in future. A 200-mile radius was chosen so that all storms known to affect any part of the island of Jamaica could be included. In addition, newspaper accounts from the *St. Christopher Weekly Advertiser* (St. Kitts) for the years 1871–1888 and 1897–1898 (no papers were available for 1889–1896) indicate that comments on disturbed weather were made for every tropical cyclone in the HURDAT record for these years that passed within an approximately 200-mile radius of St. Christopher. Even weaker events sometimes received press notice, and since they are not included in HURDAT, may represent at least tropical depressions or vigorous tropical waves.

Table 23, which is based on new tropical cyclone track maps, can be compared directly to modern track maps. Although wind speed data are not available as in HURDAT, some idea of the wind speed can be estimated from the source data in a manner used by Fernández-Partágas and Diaz (1995a,b, 1996a,b, 1997, 1999). The list may not be complete (particularly when gaps in Jamaican newspapers exist), but it is the most complete series ever built for any region of the tropical Atlantic and extends farther back in time than any other tropical cyclone series.

The frequency of hurricane-force winds at Savanna-la-Mar is a function of storm strength and the distance of the storm center from the town. Figure 10, which is derived from the data in Table 23, is a time series of all tropical storms and hurricanes that passed within 200 miles of Savanna-la-Mar for the entire hurricane season from May through November for the 253-year period 1750–2002. An eleven-year running average is superimposed.

The long-term average frequency of a tropical cyclone center being located within 200 miles of Savanna-la-Mar is 0.94 per year over the last 253 years. The number has varied from zero in many years to six in 1933. Periods of unusual tropical cyclone frequency include 1777–1790, 1894–1918, and 1931–1955. The most active period was 1777–1790, when the frequency was more than double the long-term average. Periods of low frequency of tropical cyclones include 1791–1809, 1828–1866, and 1956–1994. The latter extended period of low activity exceeded is unprecedented in the past 253 years.

Virtually every tropical cyclone to pass within 200 miles of Savanna-la-Mar in the months of July through September originated from tropical waves from West Africa in the modern record. In May–June and October–November, most of the storms originate in the western Caribbean. Figure 11 shows the 253-year time series for tropical cyclones in July through September. An 11-year running average is superimposed. The long-term average is 0.47 tropical cyclones per year. Four peaks in activity stand out, one in 1781–1790; a strong and longer-lasting peak from 1810–1832;

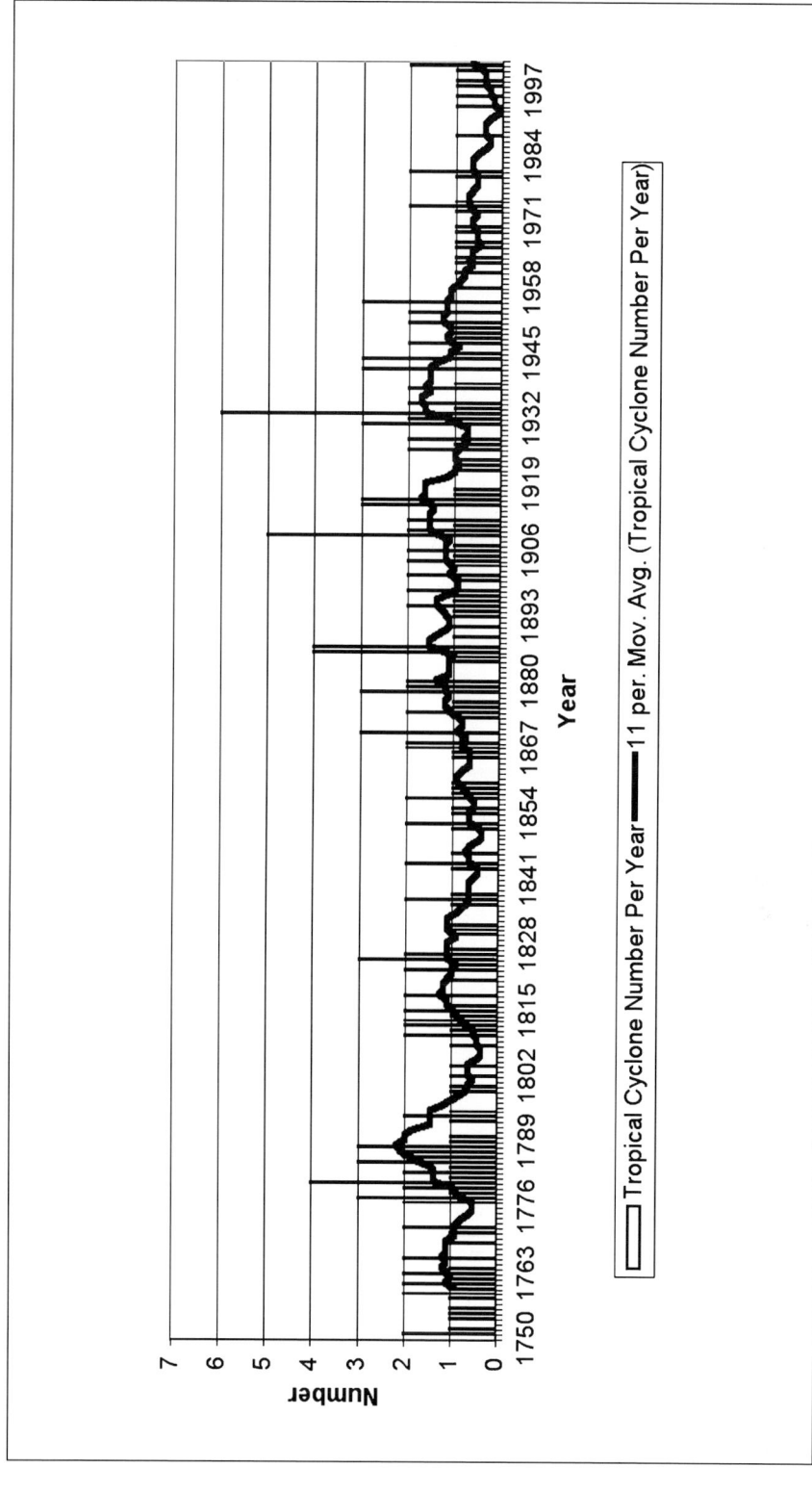

FIG. 10 Number of tropical cyclones passing within 200 miles of Savanna-la-Mar, 1750 to 2002.

TABLE 23

Dates of tropical cyclones that passed within 200 miles of Savanna-la-Mar from 1750 through 2002. The maximum sustained wind speed and the storm type (T for Tropical Storm and H for Hurricane). "None" indicates that no sustained tropical storm force or greater winds were probably felt at Savanna-la-Mar. "Gusts" indicates that while sustained wind speeds did not reach tropical storm force, some gusts probably reached and exceeded tropical storm force. All data, except for the final column, for 1871–2002 are from the HURDAT dataset. Data for 1851–1870 is a merger of HURDAT and new data collected by the author. Data prior to 1870 is a new compilation produced by the author.*

Year	Date(s)	Storm Type	Storm Type at Savanna-la-Mar	Year	Date(s)	Storm Type	Storm Type at Savanna-la-Mar
1751	23–Sep	H	H	1788	8–9 Sep	T	T
1751	6–7 Oct	T	Unknown	1788	3–Oct	T	None
1752	26–27 Sep	T	T	1789	4–Aug	T	None
1754	19–Sep	H	Gusts	1790	1–Sep	H	H
1755	8–Oct	T	T	1791	28–Sep	T	None
1756	16–17 Sep	H	Gusts	1793	21–Oct	T	None
1758	18–Oct	T	Gusts	1794	28–May	T	None
1759	14–Sep	H	Gusts	1794	26–27 Jun	T	None
1759	20–21 Oct	T	T	1795	10–11 Aug	T	None
1760	1–2 Oct	H	T	1799	2–4 Jun	T	None
1761	22–23 Sep	T	T	1800	1–2 Nov	H	T
1761	19–20 Oct	T	Gusts	1802	6–10 Oct	H	None
1762	4–5 Oct	T	None	1804	18–19 Aug	H	T
1763	16–Jun	T	Gusts	1808	27–Nov	T	Gusts
1763	5–6 Nov	T	None	1810	30–31 Jul	H	T
1764	3–4 Oct	T	T	1810	21–Oct	T	None
1766	17–Aug	H	Gusts	1811	20–Oct	H	Gusts
1766	19–Oct	H	Gusts	1812	15–16 Aug	T	None
1769	9–Aug	H	None	1812	30–Aug	T	None
1771	23–24 May	T	Gusts	1812	12–13 Oct	H	H
1772	4–5 Aug	H	None	1813	31 Jul–1Aug	H	H
1772	4–Sep	H	Gusts	1813	29–Aug	H	T
1777	30–Oct	T	None	1814	24–25 Jul	T	None
1777	21–23 Nov	T	None	1815	6–Sep	H	T
1778	5–6 Jun	T	Gusts	1815	18–19 Oct	H	T
1778	15–16 Sep	H	T	1816	1–2 Jun	T	None
1778	6–Oct	T	Gusts	1818	12–13 Oct	H	None
1779	25–26 May	T	Gusts	1818	10–11 Nov	H	H
1780	3–Oct	H	H	1821	6–Sep	H	None
1780	16–17 Oct	T	Gusts	1823	10–Jul	T	None
1781	1–2 Aug	H	H	1823	2–3 Aug	T	None
1781	6–7 Sep	T	T	1824	27–Sep	H	Gusts
1781	9–Oct	T	Gusts	1825	29–30 May	T	Gusts
1781	2–3 Nov	T	Gusts	1825	29–Sep	H	Gusts
1782	1–Jul	T	T	1825	4–6 Nov	H	Gusts
1783	1–3 Sep	T	None	1826	2–3 Sep	T	None
1783	4–5 Oct	H	Gusts	1826	9–12 Nov	T	T
1784	30–31 Jul	H	H	1827	20–Aug	H	T
1785	27–28 Aug	H	H	1830	8–Aug	H	T
1785	26–Oct	T	T	1831	13–14 Aug	H	None
1786	6–Jun	T	Gusts	1832	29–30 Aug	H	Gusts
1786	21–Oct	H	H	1836	2–3 Sep	T	None
1787	2–Sep	H	None	1837	26–27 Sep	H	T
1787	21–22 Sep	H	None	1837	24–25 Oct	H	T
1788	4–Jun	T					*(continued)*

*Shortly before the publication of this book, evidence was found that allowed the inclusion of a tropical storm on 28 September 1791 to Table 23. However, Figures 10 and 11 do not include this data point.

Year	Date(s)	Storm Type	Storm Type at Savanna-la-Mar	Year	Date(s)	Storm Type	Storm Type at Savanna-la-Mar
1838	20–21 May	T	Gusts	1859	2–Oct	T	None
1843	12–Jul	T	Gusts	1860	31–Jul	H	Gusts
1844	30 Sep–1 Oct	T	None	1865	10–Sep	H	Gusts
1844	2–3 Oct	H	T	1866	16–Aug	T	Gusts
1846	7–Oct	T	None	1867	14–Oct	H	T
1851	7–8 Nov	H	H	1867	11–25 Nov	H	T
1852	6–Oct	H	T	1868	5–Jun	T	None
1854	13–14 Sep	T	None	1868	4–5 Oct	H	T
1855	29–Aug	T	None	1870	6–Oct	H	T
1857	28–Sep	H	None	1870	16–Oct	H	T
1857	6–Oct	T	Gusts	1870	28–29 Oct	T	None
1858	22–Sep	T	None				*(continued)*

Year	Date(s)	Storm Center Max Speed	Storm Type	Storm Type at Savanna-la-Mar	Year	Date(s)	Storm Center Max Speed	Storm Type	Storm Type at Savanna-la-Mar
1873	29–30 Sep	80–40	T	None	1906	14–15 Oct	80	H	T
1874	6–Oct	80	H	T	1906	7–Nov	35	T	None
1874	1–Nov	60–80	H	T	1907	24–25 Jun	35	T	None
1875	12–13 Sep	90	H	Gusts	1909	16–17 Jul	45	T	T
1876	15–16 Oct	50	T	None	1909	6–7 Aug	50	T	None
1878	13–Aug	50	T	None	1909	23–24 Aug	90	H	H
1878	5–Sep	60	T	None	1909	15–Sep	60	T	None
1878	18–Oct	40–50	T	None	1909	8–Oct	75–85	H	Gusts
1879	3–Oct	40	T	None	1910	25–26 Aug	35–40	T	None
1879	12–Oct	50	T	T	1910	9–Sep	70	T	T
1880	6–7 Aug	80	H	T	1911	24–25 Oct	35	T	None
1880	18–19 Aug	70	T	T	1912	11–Oct	60	T	None
1884	7–Oct	50	T	None	1912	15–19 Nov	130	H	H
1885	8–Oct	40	T	None	1915	12–13 Aug	95–100	H	H
1886	27–Jun	35	T	None	1915	1–Sep	65–70	H	Gusts
1886	16–Aug	85	H	None	1915	26–Sep	90	H	Gusts
1886	20–21 Aug	85	H	H	1916	15–16 Aug	95	H	H
1886	18–Sep	85	H	T	1916	30–31 Aug	80	H	T
1887	17–May	35	T	T	1916	13–Oct	75–90	H	Gusts
1887	6–7 Aug	40–35	T	None	1917	23–Sep	95	H	T
1887	15–Sep	85	H	T	1918	4–Aug	50	T	Gusts
1887	12–Oct	60	T	None	1922	14–Oct	35	T	None
1889	15–Sep	85	H	T	1923	18–Oct	35	T	None
1891	5–Oct	45	T	None	1924	7–8 Nov	40	T	T
1893	5–Nov	35	T	None	1926	11–Sep	35	T	None
1894	23–Sep	85	H	None	1926	3–Oct	35	T	None
1895	25–Aug	85	H	T	1927	18–Oct	40	T	None
1895	19–Oct	105	H	H	1928	11–Aug	60	T	None
1896	26–Sep	105	H	H	1928	2–Sep	35	T	None
1897	13–Oct	40	T	None	1931	14–Aug	50	T	None
1898	7–Oct	50	T	None	1931	9–Sep	55	T	None
1898	30–31 Oct	50	T	T	1931	13–Sep	55	T	T
1900	3–Sep	35	T	None	1932	29–Sep	40	T	T
1901	6–Jul	55	T	None	1932	8–9 Nov	110	H	Gusts
1901	13–14 Sep	35–45	T	None	1933	2–Jul	70	H	Gusts
1903	11–Aug	95–100	H	H	1933	16–17 Jul	45	T	None
1904	13–Jun	40	T	None	1933	16–Aug	45	T	None
1904	13–Oct	35	T	None	1933	20–Sep	65	H	T
1905	4–5 Oct	70	H	None					*(continued)*

Year	Date(s)	Storm Center Max Speed	Storm Type	Storm Type at Savanna-la-Mar
1933	1–2 Oct	45–70	H	Gusts
1933	29–Oct	85	H	H
1934	19–Oct	40	T	T
1935	27–Sep	100	H	H
1935	19–Oct	60	T	T
1938	12–Aug	75	H	T
1938	23–24 Aug	85	H	None
1939	2–Nov	55	T	T
1942	25–Aug	70–80	H	None
1942	19–Sep	45	T	T
1942	13–Oct	40	T	None
1944	27–Jul	45	T	None
1944	20–Aug	105–70	H	H
1944	13–15 Oct	70	T	T
1945	12–Oct	65	H	None
1947	11–Aug	35	T	None
1947	20–Sep	35	T	T
1948	19–Sep	50	T	None
1949	12–Oct	35	T	None
1950	15–16 Oct	55	T	T
1951	17–18 Aug	75	H	T
1951	5–Sep	45–35	T	None
1953	23–Sep	40	T	T
1953	3–Oct	35	T	None
1955	23–Aug	35	T	None
1955	14–Sep	60–65	T	None
1955	26–27 Sep	130	H	Gusts
1958	2–Sep	100–70	H	None
1961	29–Oct	110	H	Gusts
1963	6–Oct	100–90	H	None
1964	25–Aug	100–70	H	Gusts
1966	30 Sep–1 Oct	100–65	H	None
1967	12–13 Sep	50	T	None
1969	31–Aug	60	T	None
1970	21–May	70–55	T	None
1973	18–Oct	40–45	T	None
1974	31–Aug	75–80	H	T
1974	16–Sep	35	T	None
1975	18–19 Sep	40–35	T	None
1980	5–Aug	115	H	H
1981	7–May	40	T	None
1981	5–Nov	60	T	None
1988	12–Sep	110–115	H	H
1994	12–13 Nov	35	T	T
1996	24–Nov	35	T	None
1998	25–Oct	115	H	Gusts
1999	14–15 Nov	55–70	T	None
2001	7–Oct	75	H	T
2002	18–Sep	45	T	None
2002	29–Sep	50	T	None

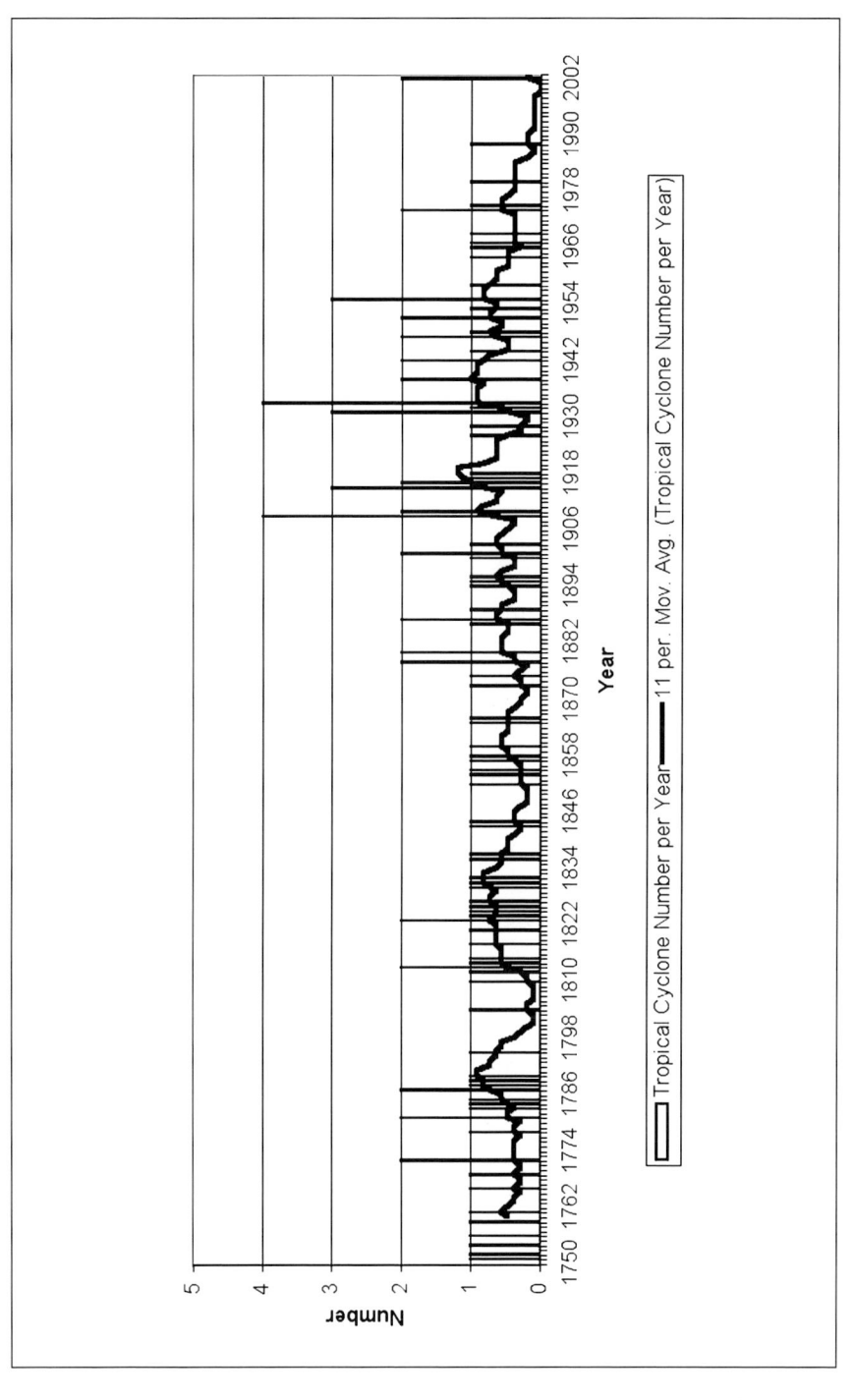

FIG. 11 Number of tropical cyclones in the months of July through September passing within 200 miles of Savanna-la-Mar, 1750 to 2002.

1909–1918; and 1931–1955. Minimum activity occurs in 1757–1780, 1791–1809, 1838–1877, 1919–1930, and 1970–2001. This latter period of little activity is longer and slightly more severe than the 1791–1809 minimum. Note the roughly sixty-year periodicity of tropical cyclone minimum activity (about 1800, 1860, 1925, 1985).

Two factors are most likely contributing to the Jamaican tropical cyclone history in the months of July through September. First, the numbers of tropical cyclones near Jamaica are partly influenced by the number of tropical cyclones that develop off the western coast of Africa and move west and northwest toward the Caribbean. If the numbers are above or below normal, then there are more or fewer opportunities for these storms to eventually reach the area of Jamaica.

A second reason for the observed variability in the Jamaican record is due to the steering currents that move tropical cyclones. Once a tropical storm or hurricane has formed, it may move away from Jamaica. The movement of tropical cyclones can be erratic, and there is an element of chance at work in the record for any particular site. As it turns out, Savanna-la-Mar is located as near as any point on land could be, to the area of the Caribbean that has been least visited by tropical cyclones in the past forty years (see Figure 15 in Landsea et al. 1999). While the entire Caribbean was visited by very few tropical cyclones in recent decades, the decline was at a maximum in the sea areas immediately west of Jamaica. This means that the low activity seen in Figure 10, although real, is slightly overstating the decline when considered on a larger regional scale.

Figure 12 is the number of tropical cyclones within 200 miles of Savanna-la-Mar for the months of May–June and October–November. (Figures 11 and 12 combined produce the results shown in Figure 10.) Unlike the months of July through September, most (but not all) tropical storms and hurricanes form near or west of Jamaica in these four months. An average of 0.47 tropical cyclones per year passes within 200 miles of Savanna-la-Mar in these months. The time series for these months show more evidence of pronounced variability in the record, particularly in the eighteenth and nineteenth centuries than in July–September. On two separate occasions, six consecutive calendar years had one or more tropical cyclones, which has not occurred again since. The most active years are 1777–1788, 1808–1826, 1867–1880, 1904–1912 and 1922–1945. Minimum activity appears from 1767–1776, 1789–1807, 1827–1866, and 1954–1993. In general, there are extended periods of near-to-above average activity from 1750–1791 and 1867–1953 and near-to-below average activity

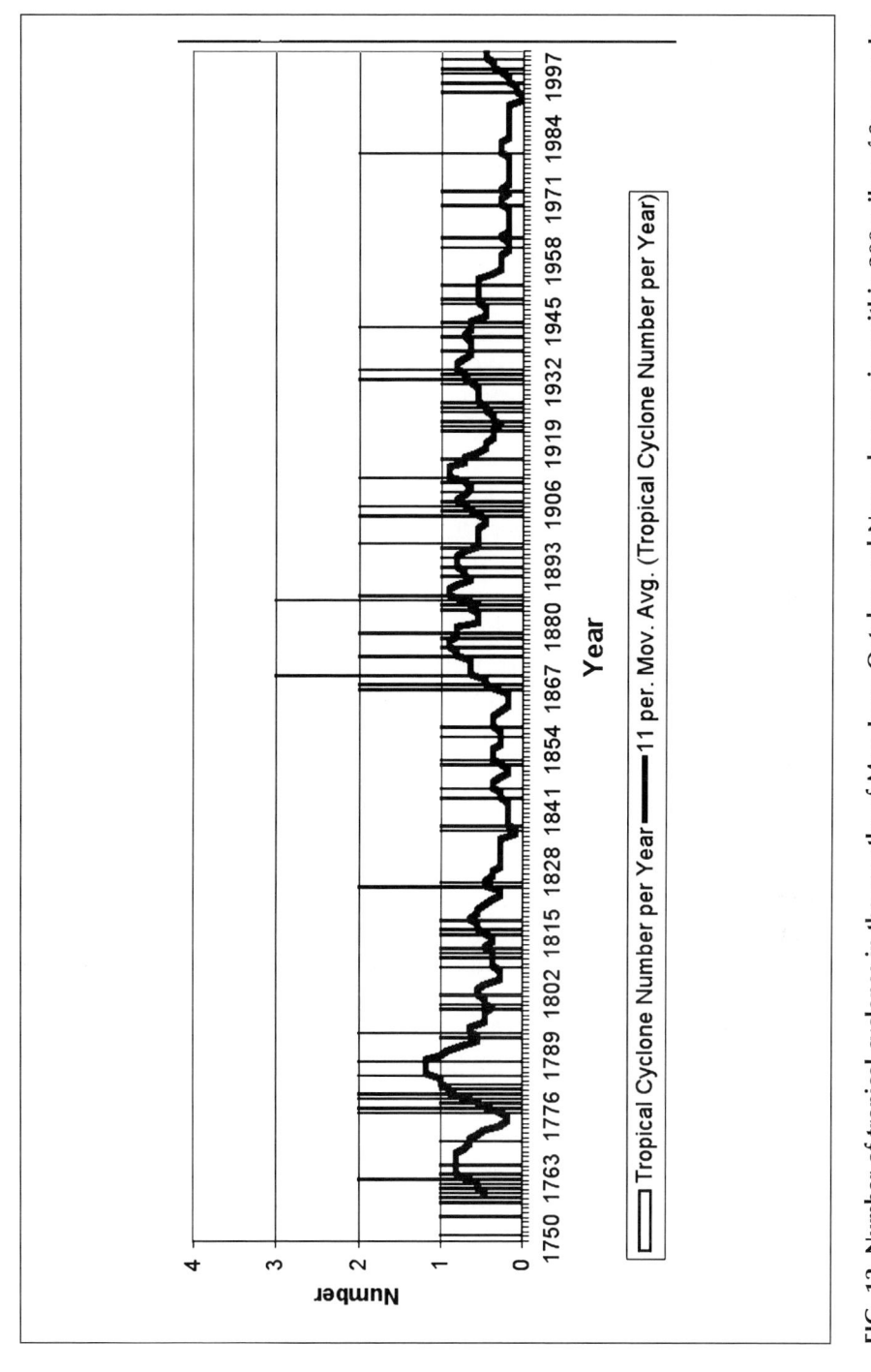

FIG. 12 Number of tropical cyclones in the months of May, June, October and November passing within 200 miles of Savanna-la-Mar, 1750 to 2002.

from 1791–1866 and 1954–2002. The late twentieth century minimum again is longer and more extreme than any other period,
but not much lower than the mid-nineteenth century levels. Only
the return of October and November tropical cyclones since 1994
has brought the eleven-year average back to normal; otherwise,
there would have been a near total absence of tropical cyclones
near Savanna-la-Mar in the last twenty years.

Hurricane winds are far more destructive than tropical storm-
force winds. Figure 13 depicts hurricane force winds felt at
Savanna-la-Mar from 1750–2002. Twenty-four documented
instances are known, in twenty-four different years. Hurricanes
have occurred in consecutive years in 1780–1781, 1812–1813,
1895–1896, and 1915–1916. The longest period between hurricanes is thirty-six years (between 1944 and 1980) and thirty-five
years (between 1851 and 1886). Hurricanes are known to have
affected Savanna-la-Mar in 1722 and 1744, so the longest gap in
the eighteenth century was twenty-nine years (between 1751 and
1780). A minimum of twenty-four hurricane days in 280 years
gives a return period of once every 11.7 years. The fifty-year
period from 1895–1944 brought eleven hurricanes to Savanna-la-
Mar, which easily exceeds the frequency at any other time. Only
three hurricanes have occurred since 1936 (a sixty-six-year
period) and this almost equals the sixty-eight-year period between
1818 and 1886. These results indicate that hurricane-force winds
at Savanna-la-Mar are approximately constant in number on the
century time scale, but the frequency of hurricane-force winds on
shorter time scales is highly variable, with several hurricanes in a
single decade, but also featuring decades of relative inactivity.

Figure 13 depicts the number of hurricanes within 200 miles of
Savanna-la-Mar by month for the period 1750–1870 compared
with 1871–2002. Prior to 1870, hurricanes were most numerous in
October, whereas August has been the most frequent month since
1871. However, the numbers are not so greatly different to ascribe
any significance other than to state that they are broadly similar the
last 133 years. The four July hurricanes prior to 1871 were all felt
on either or both the 30th and 31st of July on each occasion.

LONG-TERM TROPICAL CYCLONE VARIABILITY AS AN
INDICATOR OF LARGER-SCALE CLIMATIC VARIABILITY

What can the Thistlewood record, and other supporting data gathered with it, tell us about long-term climate variability in the
Caribbean and beyond? During two of the coldest periods of the

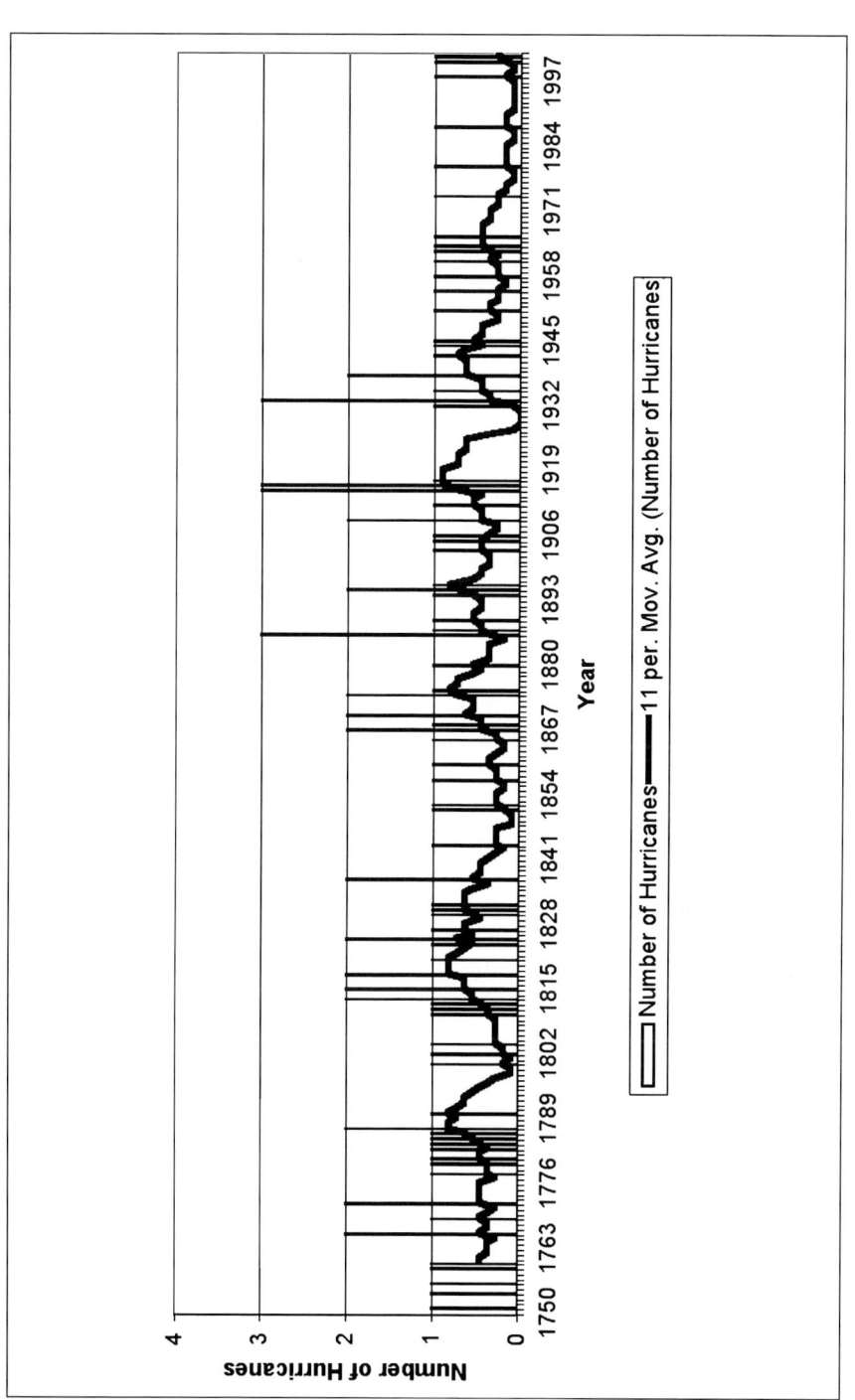

FIG. 13 Number of hurricanes only passing within 200 miles of Savanna-la-Mar, 1750 to 2002.

last 250 years in the Northern Hemisphere (1780s and 1810s), there were unusually high numbers of tropical cyclones in the Jamaican region. Another peak in activity in the 1910s also corresponds with the coldest years of the twentieth century (Jones et al. 1999; Bradley and Jones 1992; Bottomley et al 1990). Such periods are known for a weakening of the middle- and high-latitude circulation patterns and periods of unusual weather patterns at higher latitudes (Lamb 1977; Bradley and Jones 1992). Also of interest, is the indication of multi-decadal variability in tropical cyclone frequency, most notably the sixty-to-seventy year quasi-periodic numbers. These numbers and their timing are possibly associated with large-scale patterns of sea surface temperature patterns in the Atlantic Ocean (Hansen and Bezdek 1996). Long-term variations in the "thermohaline circulation" of the world's oceans (deep-ocean mixing of heat and salt) are hypothesized to produce these patterns. This pattern is most apparent in the July–September tropical cyclone numbers (see Figure 11).

Finally, the dramatic late twentieth century minimum in tropical cyclone frequency is documented in modern research. Tropical cyclone formation was greatly diminished from 1968 to 1994 in the main formation area between 10°N and 20°N (Goldenberg et al. 2001). This period coincided with severe drought in the Sahel region of West Africa (Le Barbe et al. 2002). The relatively inactive period in Jamaica pre-dates the Sahelian drought, but its most intense and record-breaking period was from 1972–1994 (Le Barbe et al. 2002; Landsea et al. 1999). The reduced numbers of Jamaican region tropical cyclone activity in the twentieth century was partly due to large-scale changes in the oceanic and atmospheric dynamics that allow the formation of tropical cyclones in the first place, and to changes in formation areas and storm tracks that are, in part, due to chance.

Because Jamaican tropical cyclone numbers are influenced by two source regions (tropical Atlantic west of Africa and western Caribbean), there may be two or more major influences that affect the numbers. Sahelian rainfall is statistically correlated with Atlantic Ocean sea surface temperatures (SST) (Folland et al. 1986). Similarly, the SST influences on Trade Wind strength and changes in the vertical shear that are tied to Sahelian drought also affect the number and intensity of tropical cyclones.

The El Niño/Southern Oscillation (ENSO) is known to have a major effect on the number of tropical cyclones forming in the eastern tropical Atlantic (Goldenberg and Shapiro 1996) and on the number of hurricanes that make landfall in the United States (Gray 1984). Warm ENSO events, which suppress Atlantic hurricane numbers, were unusually frequent in the late twentieth century (Allan 2000). Similarly, the strength of the Atlantic Trade Winds and atmospheric pressure are correlated with tropical cyclone numbers and intensity. When the Trade Winds are below (above) average strength and pressure below (above) average, then the number of tropical cyclones and hurricanes tends to be above (below) average (Shapiro 1982). However, this correlation is not apparent in 1750–1786 in Jamaica, but since the record is only for a small part of the tropical Atlantic, it is likely that a much larger area needs to be sampled.

The long-term perspective provided suggests that the most recent decades are anomalous with respect to the 253-year record in the Jamaican region. If indicative of the entire North Atlantic basin, then this would suggest that the atmospheric and oceanic conditions that favor tropical cyclone formation in the Atlantic were highly anomalous in the late twentieth century. Meridionally antisymmetric differences in global SST anomalies between the Northern and Southern Hemisphere (cooler Northern Hemisphere/warmer Southern Hemisphere oceans) in the observed record are associated with reduced tropical cyclone numbers (Landsea et al. 1999). This is a pattern of secular SST changes that are also associated with significant rainfall anomalies over the Sahel (Folland et al. 1986). The Atlantic SST variability is also influenced by ENSO-associated SST fluctuations in the Pacific Ocean that have a significant impact on rainfall in the Caribbean. For example, when SST anomalies are of opposite sign in the eastern tropical Pacific and the tropical North Atlantic (cold Pacific/warm Atlantic), then rainfall in the Caribbean region is enhanced (Enfield and Mestas-Nuñez 2000). The Thistlewood record, which suggests rainfall amounts equal to the twentieth century, but with considerably lower temperatures, provides evidence that this pattern of SST anomalies may have prevailed in the eighteenth century. Also, the late twentieth century featured more frequent warm ENSO than cold ENSO events, which may indicate that warm ENSO events were fewer and weaker in previous periods of high tropical cyclone activity.

The Thistlewood record highlights the significant decline of Caribbean tropical cyclones in recent decades. Global warming may be a culprit in this change. Warmer temperatures, along with drier conditions in Jamaica in recent decades (Chen et al. 1997), in part, due to weaker weather disturbances and fewer tropical cyclones, are making the modern climate of Jamaica increasingly remote from that experienced, and so carefully documented, by Thomas Thistlewood.

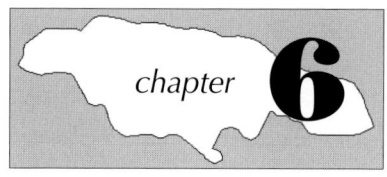

chapter **6**

CONCLUSION

The Thistlewood weather record allows the first detailed picture of eighteenth century Caribbean climate. His record is the most important and complete weather record from anywhere in the global Tropics for such an early time. Located in a region sensitive to some of the most important oceanic and atmospheric elements that drive regional and global climate. it immediately sets limits as to the possible deviation and magnitude of climatic anomalies relative to the modern climate. This high-quality record establishes that the Caribbean was about 2°F cooler than in 1951–1980 (and even cooler than that in the period since 1980), yet the annual rainfall was nearly equal to that of 1921–1977. In addition, wind speed records indicate that the Azores High, during the summer months at least, was less likely to build with strength into Jamaica, indicating its persistent displacement elsewhere in the Atlantic basin, relative to the twentieth century. Based on our knowledge of weather conditions in Europe at this time, it is likely that the Azores High was displaced to the north over the north-central Atlantic Ocean, allowing more northerly winds and cooler conditions to prevail in western Europe. Given that Europe experienced colder conditions year-round, it is probably safe to assume that the Azores High was weaker and/or displaced from its twentieth century position year-round.

A weaker Azores High would allow for a northward movement of the equatorial intertropical convergence zone (the area where the Northern and Southern Hemisphere Trade Winds meet). The cooler, moister weather in Jamaica in the eighteenth century argues for such a scenario. Based on modern data, such a pattern of wind and moisture is consistent with relative sea surface temperature anomalies that make the equatorial Pacific cooler than the equatorial Atlantic. In such times, Central

America and the Caribbean receive more rainfall than normal. Rainfall is also enhanced in the Sahelian region of West Africa. Such a pattern argues for relatively stronger Trade Winds in the Pacific Ocean and relatively weaker Trade Winds in the North Atlantic Ocean.

Weaker Atlantic Trade Winds, particularly in the summer, would be favorable for tropical cyclone formation in the main formation regions. A new 253-year time series of Jamaican area tropical cyclones shows no trend in tropical cyclone numbers, although a full basin-wide census of tropical cyclones is needed to establish this. Nonetheless, the greatest drought (1972–1994) in West African history corresponds with the longest and most severe suppression of Atlantic basin hurricanes in the modern record. The 253-year Jamaican record and historical accounts from Africa (Nicholson 1980) indicate no equal to the late twentieth century hurricane minimum. An approximately sixty-year cycle in tropical cyclone numbers (see Figure 11) is the best evidence yet from historical records that such a cycle may be a real manifestation of a fundamental oceanic/climatic oscillation.

The number of tropical cyclones, and particularly hurricanes, in the tropical Atlantic is strongly influenced by the El Niño/Southern Oscillation (ENSO) phenomenon. Periods of suppressed hurricane activity are greatest in warm ENSO (El Niño) events and periods of high activity are greatest in cold ENSO (La Niña) events. The historical record of warm/cold ENSO events is incomplete, particularly prior to the nineteenth century. The Thistlewood record provides new information for increasing our knowledge of historical ENSO events, but it must be accompanied with independent data from elsewhere in the region. Based on the inferences made about the most likely prevailing state of tropical oceanic winds and temperature, the ratio of warm to cold ENSO events may have been slightly less from 1750–1786 than in the twentieth century. The very high ratio of warm to cold ENSO events in the late twentieth century coincided with reduced Atlantic hurricane activity. However, additional factors, such as the strength of the Atlantic Trade Winds, also affect Atlantic hurricane numbers. The atmospheric and oceanographic interactions that produce climate are yet to be understood on physical grounds. The Thistlewood record provides invaluable data to assist climate modelers in their reconstruction of past, current, and future climate.

ACKNOWLEDGMENTS

I would like to acknowledge my deepest appreciation to Lord and Lady Monson for granting me special privileges that greatly sped up the data acquisition and digitization process of the Thistlewood weather record. With permission from Lord and Lady Monson and the Lincolnshire Archives, I have quoted extensively from portions of the original (non-digitized) weather record. Mr. R.B. Hall, of Lincoln, kindly visited and consulted the original manuscript on my behalf when the microfilm version was difficult to discern. My thanks to my wife, Anne, for her patience and understanding during my work on this book. My research was encouraged and supported by Dr. Cary Mock, Department of Geography, University of South Carolina, Columbia, and Dr. Kam-biu Liu, Department of Geography and Anthropology, Louisiana State University, Baton Rouge. My thanks also to my reviewers who closely read the manuscript and asked very informed questions that helped to shape the final product. I have also benefited from the assistance of the always-professional staff of the American Philosophical Society. A Franklin Research Grant made by the American Philosophical Society made this work possible.

In loving memory of my brother
Lloyd Grant Chenoweth, Jr.
who died shortly before this book was published.

METADATA FOR BLACK RIVER AND SAVANNA-LA-MAR WEATHER STATIONS

Manuscript Location: Lincolnshire Archives, Lincoln, U.K. MS condition generally good with some fading of ink and water stain and damage to some originals. Legibility is generally good to excellent, except for the year 1760. Record is available on microfilm. No previously published data.

BLACK RIVER WEATHER STATION LOCATION

Placename: Vineyard Pen, St. Elizabeth Parish.
Coordinates: 18°02′ N 77°49′ W
Ground Elevation: Unknown.
Date of First Observation: 26 July 1750 (New Style)
Date of Last Observation: 21 July 1751 (New Style)
Instrumental Observations: None
Noninstrumental Observations: State of Sky, Weather, Wind Force
Time of Observation of Instrumental Observations: Not applicable.
Other Relevant Information: Located about five miles east of the port of Black River, in a low-lying marshland along the Broad River, a tributary of the Black River. The Burnt Savannah Mountains lie to the east.

FIRST SAVANNA-LA-MAR WEATHER STATION LOCATION

Placename: Egypt Plantation.
Coordinates: 18°14′ N 78°10′ W (as calculated by Thistlewood, 18°15′ N 78°05′ W; longitude calculated from "London")
Ground Elevation: Unknown, but approximately 15 feet.
Date of First Observation: 12 January 1752

106

Date of Last Observation: 3 September 1767
Instrumental Observations: Temperature and Rainfall
Time of Observation of Instrumental Observations: Temperature (Sunrise, Noon, Sunset); Rainfall (Once daily at Sunrise and the amount charged to the previous calendar day)
Location of Rain Gauge: Not stated, but assumed to be in the garden near the residence.
Elevation of Rain Gauge above Ground: Not stated.
Type of Rain Gauge: Not stated, but apparently homemade.
Date of First Rain Gauge Observation: 1 July 1760
Date of Last Rain Gauge Observation: 3 September 1767
Location of Thermometer: Not stated, but assumed to be similarly exposed as given in Bread Nut Island thermometer location.
Elevation of Thermometer Above Ground: Not stated, but assumed to be similarly exposed as given in Bread Nut Island thermometer location.
Dates of Thermometer Observations: 1 May 1753 to 24 June 1754 and 1 January 1764 to 3 September 1767, with a few sporadic observations as early as November 1761.
Manufacturer and Type of Thermometer Used: First thermometer of apparent Hauksbee manufacture. Scale calibrated into 100 equal parts by Thistlewood. Second thermometer manufactured by Bury & Edmonds, London, Fahrenheit scale.
Other Relevant Information: See Bread Nut Island Plantation for descriptions applicable to both Egypt and Bread Nut Island plantations.

SECOND SAVANNA-LA-MAR WEATHER STATION LOCATION

Placename: Bread Nut Island Plantation (formerly Paradise Pen; today known as Rock Dona (Hall 1989, Fig. 3A). Comprises about 300 acres, bounded on the west by the Cabaritta River, on the north by the property of George Goodin, "on the east by Kirkpatrick Pen and lands belonging to Goodin and Thomas Hall Esqr., on the south-east the property extended beyond the King's road and on the south it was bounded by the King's road to Savanna la Mar and lands belonging to John Prynold." Source: Douglas Hall. *In Miserable Slavery. Thomas Thistlewood in Jamaica, 1750–86.* Warwick University Caribbean Studies, New York: Macmillan Press Ltd, 1989.
Coordinates: 18°14′ N 78°10′ W (as calculated by Thistlewood, 18°15′ N 78°10′ W)

Direction and Distance from Egypt Plantation: About N 67° 1/2 E distance, 64 chains, 10 links. Value derived from compass, magnetic declination, 6 degrees East. About one mile east-north-east of Egypt Plantation House and about three miles northwest of the harbor of Savanna-la-Mar.

Ground Elevation: 16.5 feet. Thistlewood gives the elevation as 15 to 16 feet above mean high water mark of sea, which has a maximum range from low to high of 20 to 24 inches. This gives an approximate elevation of 16 to 17 feet above mean sea level.

Date of First Observation: 4 September 1767

Date of Last Observation: 15 November 1786

Instrumental Observations: Temperature and Rainfall

Time of Instrumental Observations: Temperature (Sunrise, Noon, Sunset); Rainfall (Once daily at Sunrise and the amount charged to the previous calendar day)

Location of Rain Gauge: Not stated, but assumed to be in horticultural garden area near the residence.

Elevation of Rain Gauge Above Ground: Not stated.

Type of Rain Gauge: Not stated, but apparently homemade.

Date of First Rain Gauge Observation: 4 September 1767

Date of Last Rain Gauge Observation: 15 November 1786

Location of Thermometer: On a north-facing earthen wall for noon and sunset readings. Exposed so that it is never exposed to direct sunlight and with a free access of air. After the sunset observation, the thermometer was moved to a well-ventilated location under an eave of the main house and remained in this location until the sunrise observation was taken.

Elevation of Thermometer above Ground: About six feet.

Other Relevant Information: "Both Egypt & Breadnut Island are very dry weather places, although almost surrounded by deep Marshes and have often great Floods of Water pass thro' them, occasioned by the rain which falls up in the country. The mountains surrounding us almost in form of a half moon, from nearly SE to NW, at the distance of 5, 6, or 7 miles, and in some places more. To the south is the sea, but the sight of it is interrupted by the tall Mangroves which grow in the marshes." (Thistlewood, Monson Collection, 31/89, pp. 75–76)

". . . .Egypt dwelling house and the house where I now live at Breadnut Island, are both about a mile from the sea, and enclosed almost on every side with tall Mangroves, in some places at the distance of 2 or 3 hundred yards only, but in others 7 or 8 hundred." (Thistlewood, Monson Collection, 31/89, pp. 75–76)

APPENDIX 2

DEFINITIONS OF COLUMN "HD" USED IN DIGITIZED DATABASE OF THISTLEWOOD WEATHER JOURNAL

Time of Day (TOD) Modifier	Encoded Integer
Soon after (stated time in column HH)	1
A little before (stated time in column HH)	2
Afterwards (next time sequence following time in column HH)	3
Indicates that a specific time or range of times appear in columns "FROM" and "TO"	4

PRESENT WEATHER (WW1 AND WW2) TYPES USED
BY THISTLEWOOD AND THEIR CORRESPONDING
ENCODED INTEGER VALUES USED IN THE
DATABASE OF THISTLEWOOD WEATHER
OBSERVATIONS

Definition	Encoded Integer
Solar halo	0
Lunar halo	1
Mock sun	2
Somewhat hazy	4
Haze or hazy weather	5
Excessive haze	6
Earthquake	9
Mist	10
Whirlwind	12
Ugly or threatening sky	13
Somewhat squally	14
Hard squalls of wind	15
Excessive hard squalls	16
Waterspout	17
Sea roars	18
A "North"	20
Very drying winds	30
Foggy	40
Thick fog	45
Very thick fog	46
Lightning	90
Thunder	91
Thunder and lightning	92
Thunder at a distance	93
Thunder far off	94
Thunder remote	95
Lightning and thunder at a distance	96
Lightning and thunder far off	97
Lightning and thunder remote	98
"Severe" thunderstorm	99

CANDIDATE TROPICAL STORM AND HURRICANE
EVENTS FROM THE THISTLEWOOD RECORD
COMPARED WITH INDEPENDENT
CONTEMPORARY DATA AND WITH PREVIOUS
PUBLISHED COMPILATIONS.

All storm dates are converted to the modern Gregorian Calendar. Contemporary newspaper dates and previously published dates are unchanged.

1. THISTLEWOOD RECORD, MONSON COLLECTION 31/2, 21–22 SEPTEMBER 1751

Equivalent or related storms in Millás (1968): Cases 56 and 57, pp. 207–209.

Equivalent or related storms in Tannehill ([1938] 1952): (a) 1751, September 15 Santo Domingo; (b) 1751, September 21–22, Santo Domingo; (c) 1751, October, Jamaica, Santo Domingo.

Millás (1968) believed that this storm passed to the south of Jamaica and that a second, more severe hurricane passed near Hispanolia and to the north of Jamaica in October. He relied on the accuracy of Edward Long's dismissal of the storm as even being of hurricane strength and other poorly dated manuscripts for his conclusion. Millás also relied on an account of English ships lost in a hurricane from *The Gentleman's Magazine* that gave no details about the location or date the ships were lost.

This storm was a major hurricane and can be traced back to Antigua on 19 September, where twenty-four ships were driven on shore and the storm described as " . . .the most violent Hurricane that has been known here in the Memory of the oldest Man living" (*Pennsylvania Gazette*, 24 October 1751). It was undoubtedly the same hurricane that Millás attributes to the month of October that struck the southern sections of Hispanolia.

Thistlewood wrote in his diary a large list of all the ships that were sunk, driven ashore, or otherwise damaged at Port Royal, Jamaica.[32] His own account of the storm, which thirty years later he considered to be second only to the great hurricane of October 3, 1780, noted that it caused considerable damage. There was fore-warning of the storm the previous day.

Tuesday 10th [21st New Style]. Moderate Breezes Northerly, at times Cloudy. Old Mr. Hopkins says this timely North is a sign that the Heavy Rains [the Rainy Season] are over. But Mr. Dorrill says he never saw a North out of Season, but was followed by excessive bad weather.

Wednesday 11th [22nd New Style]. Dark & Cloudy. About 6 in ye Morning began to blow very fresh at NNE, continued with drizzling rain. About 11 blew harder at NE, and at noon blew a Storm. The hard gusts raised ye Sea Water up like mist or smoke. From 3 P.M. to Sunset blew a hurricane, attended with rain, [wind from] NE to E by S. Blew ye shingles off ye stables and building house and all ye thatch off ye cooperage and Trash house. Burst open ye great house windows, that were secured by strong board, blew ye weather boards and ridge boards off ye great house & every room full of water. The Peppermint Tree by ye Tombs blown down and many of ye Orange trees out of the rows before the great house. All the Pomengranates [sic] in the garden blown down, and every thing tore to pieces almost. Plantain trees are perfectly down. About Sunset all the white people . . . sheltered in the storehouse and hurricane house. About 11 P.M. returned to the great house, went to bed and laid down, the wind rather abated. By times in the night blew very hard as ever almost [always] attended with hard rain and lightning but no thunder. [Note: The wind probably blew so strong that thunder could not be heard.]

Thursday 12th [23rd New Style]. Continues to blow very hard with hard rain. About 9 A.M. [the] wind came to SSW but presently went back to E by S. Towards Noon more moderate with fair weather by intervals. About Noon I walked down the [Sea Wall] up to the knees in water. Strange havoc there, Mordiner's house quite down, all ye South wall (except about 11 feet) of the New store house washed clean away, and some of the east and west walls pretty much there . . . all the heavy timber floated long way up in a surprising manner. The boards, staves, and shingles blown about as if they were feathers. Most of the new wharf washed away, vast wreaths of sea weeds drove a long way upon the road. A heavy gun roller case carried a long way from where it lay, and half-buried in

32 Thistlewood, Monson Collection, 31/2, pp. 218–223.

the sand. Mr. Dorrill says the sea has raged with greater fury than either in the years [17]22 or [17]44.

Friday, 13th [24th New Style]. Fresh breezes, continual thunder to the Southward. Walked a long way by the sea side, each side of the [sea wall], no birds stirring, shot 2 snipes. Many fish thrown up dead upon the shore… The Mangroves along the Sea side twisted to pieces and few leaves left on them, all the woods and mountains look open . . .the woods appear like our woods in England in the fall of the leaf when about half down (Thistlewood, Monson Collection, 31/2).

The hurricane damage (denuded landscape, heavy objects carried well inland, sea inundation) is consistent with a major hurricane. This storm was felt in the Florida Keys on 26 September (*South Carolina Gazette*, 7 October 1751) and may have been the same storm that destroyed St. Mark's, Florida that year (*South Carolina Gazette*, 18 November 1751). This powerful and extensive storm should be included among the more severe hurricanes of the eighteenth century.

2. THISTLEWOOD RECORD, MONSON COLLECTION 31/39, 26–28 SEPTEMBER 1752

Equivalent or related storms in Tannehill ([1938] 1952): (a) 1752, September 15. Charleston [South Carolina]; (b) 1752, [n.d.] September, Charleston [South Carolina]. Two hurricanes at Charleston in one month.

Charleston, South Carolina was devastated by the most serious hurricane in its recorded history on 15 September 1752. A second storm hit Charleston on 1 October. The origin of this second storm can be tracked back to Jamaica. The weather became disturbed at Savanna-la-Mar on 24 September, with moderate breezes from a southerly direction. On 25 September, toward morning, there was a hard squall of wind and rain from the south. On the night of 25–26 September the weather deteriorated with "some hard squalls of wind and rain from the south" and the day characterized as "cloudy with frequent squalls." On 27 September, the poor weather continued. "Fresh gales Southerly with frequent squalls of wind and almost continual rain. At times thunder far off. Continued to blow very fresh in the night." On 28 September, the fresh southerly gales continued in the morning but by afternoon the wind came around more to the southeast and the squalls ceased.

This storm apparently gathered strength as it slowly moved north and passed west of Jamaica. On 26 September, the *Griffin* frigate was driven ashore on the Isle of Pines, Cuba by the storm and "a most terrible Hurricane was felt at the Havana, in which every vessel was drove ashore" (*Boston News Letter*, 11 January 1753). This same storm continued to move north and passed near Charleston.

3. THISTLEWOOD RECORD, MONSON COLLECTION 31/41, 18 SEPTEMBER 1754

Equivalent or related storm in Millás (1968): Case 58, pp. 209–210.

Equivalent or related storm in Tannehill ([1938] 1952): 1754, September (13–15?), Leeward Islands. Santo Domingo.

This documented but questionably dated storm is listed in most previous compilations but its effects in Jamaica were not mentioned. This hurricane affected Antigua (*Pennsylvania Gazette* 31 October 1754) on 13 and 14 September, and by 4 P.M. on 17 September the slow-moving hurricane was met by a ship between Haiti and Jamaica. The storm lasted 26 hours (*Pennsylvania Gazette*, 24 October 1754). The storm was felt at Port Royal, Jamaica on 18 and 19 September as it passed to the north of Jamaica (*Boston News Letter*, 21 November 1854; HMS *Hind* logbook, ADM 51 455; HMS *Southampton* logbook, ADM 51 914). At Egypt Plantation, Thistlewood had daytime winds that were moderate and from the North (instead of the normal Southerly sea breeze). The next morning he had frequent squalls of wind from the North with rain and that night "a many very hard squalls of wind and rain." On 19 September, the squalls were less severe and frequent and variable between North and West. The ultimate fate of this storm is not known, but if its slow rate of movement continued, then it may be the same storm that was felt off North Carolina on 25 and 26 September (*Pennsylvania Gazette*, 10 October 1754).

4. THISTLEWOOD RECORD, MONSON COLLECTION 31/41, 7 OCTOBER 1755

This previously undocumented tropical storm affected both Port Royal and Savanna-la-Mar. Hard squalls and unsettled weather on 7 October were noted in the logbook of HMS *Severn* (ADM 51 984). Thistlewood provided more details of its effects at Egypt Plantation. Southerly winds were first recorded on 6 October. On 7 October he wrote, "From Midnight till Noon continual hard showers of rain with hard squalls of wind from the Southward & Thunder. P.M.

most part hard rain with thunder and lightning, but not quite such hard squalls as in the forenoon. Also at night some showers."

This slow-moving storm produced South and Southwest winds at Egypt Plantation through 11 October. On 11 October, Thistlewood made a comment on the effects of the rains, which is very rare in his weather diary, tersely stating "a great flood." No further accounts of this storm elsewhere in the region have been uncovered. The tropical storm developed in the northwest Caribbean Sea and passed nearest, and perhaps reached its greatest strength, on 7 October to the west of Jamaica.

5. THISTLEWOOD RECORD, MONSON COLLECTION 31/42, 16–17 SEPTEMBER 1756

Equivalent or related storm in Millás (1968): Case 61, pp. 211–212.

Equivalent or related storm in Tannehill ([1938] 1952): 1756, September 12, Martinique.

This storm appears to be the same storm that was felt in the Leeward Islands on 12 September (Millás 1968; *Boston News Letter*, 11 November 1756; *Pennsylvania Gazette*, 25 November 1756). If so, the storm passed to the south of Jamaica and then perhaps turned to the northwest while still south of the island.

> Thursday 16th: From Midnight till almost Noon strong gales, most part hard rain with lightning and loud thunder. P.M. more moderate. In the evening drizzling rain, lightning and thunder at a distance. Southerly winds ever since Midnight.

> Friday 17th: In the night some hard squalls of wind and rain. From Sunrise till Night more moderate but frequent squalls of wind and rain with thunder far off.

6. THISTLEWOOD RECORD, MONSON COLLECTION 31/44, 16–17 OCTOBER 1758

The 1758 hurricane season was active, but Jamaica escaped storms of significant intensity. This October storm probably originated in the western Caribbean Sea and moved to the north.

> Monday 16th. From Midnight till Noon moderate breezes Southerly with some showers of rain, thunder remote. P.M. Fresh gales, cloudy, some showers of rain. At night, excessive hard squalls of wind and rain from the South.

Tuesday 17th. Till Morning hard squalls of wind and rain from the South. Continues fresh gales, cloudy and somewhat squally, thunder remote. P.M. more moderate, but cloudy, with variable winds. At night, lightning.

This storm is probably the same storm that passed near St. Augustine on an unspecified date prior to 27 October (*South Carolina Gazette*, 2 November 1758). The storm appears to have moved along the eastern U.S. coast as a storm from the southeast drove a ship ashore and destroyed it on 24 October; this storm was also felt at New York City (*Boston News Letter*, 16 November 1758). A "violent gale" was also felt on an unspecified date in late October on Cape Breton Island, Nova Scotia, with several ships driven ashore and totally lost (*Boston News Letter*, 14 December 1758).

7. THISTLEWOOD RECORD, MONSON COLLECTION 31/45, 13 SEPTEMBER 1759

Equivalent or related storm in Millás (1968): Case 66, pp. 214–215.

Equivalent or related storm in Tannehill ([1938] 1952): 1759, September, Gulf of Mexico, Florida.

Equivalent or related storm in Ludlum (1963): 1759, September, S.W. Florida.

This tropical storm first appears to the south of Jamaica on 12 September. The day at Savanna-la-Mar began with light winds but "about 10 A.M. sprang up a fresh gale about SE with frequent hard squalls of wind and rain which continued all the afternoon and night" and the sky displaying "a very furious look." The squalls (some hard) continued the next day, but the rain ceased on this day.

This storm may be the same one in Millás (1968), based on Southey's work that affected central Cuba sometime in September 1759. Ludlum (1963) also mentions a storm in southwest Florida sometime that same month. If so, this storm passed near western Jamaica on a northerly course and strengthened into a hurricane as it passed over Cuba and the Straits of Florida.

8. THISTLEWOOD RECORD, MONSON COLLECTION 31/45, 20 OCTOBER 1759

This previously undocumented storm brought tropical storm force winds to western Jamaica on the night of 20–21 October.

Saturday 20th. Light winds and almost continual moderate rain till about 9 A.M. Then clear[ed] up somewhat but about an hour after sprung up a moderate gale which continued with frequent showers of rain, thunder at a distance. At night, hard squalls of wind (Southerly) and rain, with lightning and loud thunder.

Sunday 21ˢᵗ. Frequent squalls of wind and rain till morning, then fresh gales, cloudy and somewhat squally. P.M. Cloudy. At night, light winds, fair weather.

At this late time of the hurricane season, tropical cyclones move between north and east from their most frequent breeding ground in the northwest Caribbean. The storm of 20–21 October may be related to a gale experienced to the south of Charleston, South Carolina on 25 October. According to the *South Carolina Gazette* of 1 November 1759, "The Sloop *Nancy & Becky* . . . bound for Jamaica, sail'd over the bar the 24ᵗʰ [October] but meeting with a hard gale of wind in the Gulph [Stream] the day following, she lost her bowsprit and sprung a leak, which obliged her [to return to Charleston]."

9. THISTLEWOOD RECORD, MONSON COLLECTION 31/46, 1–2 OCTOBER 1760

The next tropical cyclone to be felt at Savanna-la-Mar also made an impact on the weather in South Carolina. The latter part of September and early October brought a sustained period of Southerly winds. By the evening of 1 October "very hard squalls of wind and rain, variable from South to West" were felt at Egypt Plantation. "From Midnight till morning (2 October) frequent strong blasts of wind from the South with several minutes almost calm between the blasts & cloudy. [P.M.] Continues fresh gales Southerly with some squalls of wind and rain, thunder at a distance, also lightning." Fresh Southerly gales continued the next three days but the worst weather was over.

This storm moved to the north and northeast and was felt by ships at sea off the coast of South Carolina (*South Carolina Gazette*, 25 October 1760) and near Cape Hatteras and further to the northeast (*Pennsylvania Gazette*, 30 October 1760).

10. THISTLEWOOD RECORD, MONSON COLLECTION 31/47, 22–23 SEPTEMBER 1761

An area of low pressure appears to have formed near Jamaica on 19 September, moved to the west, and deepened west of Jamaica.

No additional evidence has yet been uncovered that the storm passed over Cuba or the well-traveled shipping routes in the region. The period from 1756–1763 covers the Seven Years' War between France and England and newspaper accounts of tropical weather are less frequent at this time. The daily sequence of weather suggests that the storm may have meandered west of Jamaica while drifting slowly away from the island.

> Friday 18th September 1761. Moderate breezes Southerly and at times cloudy."

> Saturday 19th. Moderate breezes variable from NE to E and then to South and SW & somewhat cloudy. P.M. Cloudy, thunder remote. At night, lightning.

The prevailing southerly winds from 15–18 September briefly gave way to NE winds, followed by SW winds. This suggests a weak area of low pressure formed to the southeast of Savanna-la-Mar and passed over Egypt Plantation as an incipient tropical low.

> Sunday 20th. Calm, fair and very hot. Afterwards light winds and at times cloudy. In the evening, cloudy, some few drops of rain, thunder at a distance. At night, lightning.

The calm winds and apparent hot temperatures are typical comments when a tropical storm or hurricane is developing or already formed. On 21 September, a moderate southerly breeze blew and they freshened to a fresh breeze on 22 September. On the evening of the 22nd, a moderate southerly breeze blew with "an abundance of lightning all round, thunder remote. About 11 P.M. began to blow hard squalls of wind from the SE."

> Wednesday, 23rd. Continued hard squalls of wind from the South East till about 2 A.M. with a gloomy dark sky, an abundance of lightning and thunder far off. Morning, light winds variable and cloudy. Afterwards fresh breezes SE, somewhat cloudy and hazy but extreme drying. P.M. thunder remote. At night, lightning.

11. THISTLEWOOD RECORD, MONSON COLLECTION 31/47, 19–20 OCTOBER 1761

This previously undocumented tropical cyclone passed west and north of Jamaica and was felt in western Hispanolia on 22 October (*Boston News Letter*, 14 January 1762; *Pennsylvania Gazette*, 14 January 1762).

Monday 19th. Moderate breezes Southerly, cloudy, thunder remote. About 11 A.M. came on a very hard squall of wind and rain from the SW with terrible thunder and lightning. P.M. Fresh breezes SE and cloudy, some squalls of wind and rain, thunder at a distance. At night, very hard squalls of wind and rain from the South with thunder and lightning.

Tuesday 20th. Fresh gales squally from the SW, thunder at a distance. P.M. moderate gales Westerly, some squalls of wind and rain, thunder remote. At night, lightning, and light airs variable, cloudy.

Wednesday, 21st. Light airs variable, cloudy, at times drizzling rain. Towards Noon moderate breezes North, most part cloudy.

The wind sequence suggests that a weakening cold front pushed into western Jamaica during the 21st. Along with the stormy weather experienced on the 22nd in western Hispanolia, this suggests that the tropical storm moved almost due east after passing to the northwest of Jamaica. It probably continued east or east-northeast over the southern Bahamas and probably weakened soon after.

12. THISTLEWOOD RECORD, MONSON COLLECTION 31/49, 15 JUNE 1763

A period of increasingly disturbed weather prevailed at Egypt Plantation from 10–14 June 1763. By 14 June, disturbed weather was also noted at Spanish Town, where Long recorded cloudy and showery weather all day with a strong sea breeze, followed by its continuation with rain and squalls during the night of 14–15 June (Long Collection, Add. MS 18963 ff. 1–24v).

Thistlewood had frequent hard squalls of wind all morning on 15 June and more moderate weather in the afternoon, almost a complete reversal of the normal diurnal pattern of the weather. Winds remained south to southeast and blew moderate breezes to fresh gales through 21 June, as the center of low pressure remained in the northwestern Caribbean. No additional data has been found to allow this storm to be tracked further to the north.

13. THISTLEWOOD RECORD, MONSON COLLECTION 31/50, 2–3 OCTOBER 1764

The first violent storm of the 1760s to affect Savanna-la-Mar began on the night of 2–3 October. Hard squalls of wind and rain blew from the southeast quarter. The weather deteriorated on the 3rd.

Wednesday 3ʳᵈ. Frequent squalls of wind and rain from SE, but variable a few points on either side. [A.M.] Continues very dark & black. P.M. Hard squalls of wind and rain from SE, increasing till past Midnight, when the squalls of wind were not quite so strong, but violent rain. About 2 A.M. lightning and thunder at a distance, which (I) heard several times. Pluv[iometer] — 7.64 inches, viz. From sunrise to setting 2.43 inches and from Sunset to Sunrise Thursday morning 5.21 inches.

Thursday 4ᵗʰ. Continue hard squalls of wind and rain at SE, thunder remote. P.M. rather more moderate. At night, moderate with some showers of rain, thunder at a distance. Pluv[iometer] 2.10 inches".[33]

There is no further indication of this storm, so it most likely pursued a northwest or west track toward the Yucatan Peninsula and the western Gulf of Mexico.

14. THISTLEWOOD RECORD, MONSON COLLECTION 31/52, 16 AUGUST 1766

Equivalent or related storm in Millás (1968): Case 74, pp. 219–220. Case 75, p. 220.

Equivalent or related storm in Tannehill ([1938] 1952): (a) 1766, August 13, Martinique; (b) 1766, August 16, Jamaica. Probably same storm as preceding.

A powerful hurricane brushed by Jamaica on 16 August 1766. During the afternoon, from his home one mile from the coast, Thistlewood "heard the sea roar prodigiously" with fresh south gales "increasing & variable to East & ESE, at times small rain and a dreadful roaring of the sea, or dismal gale. About 10 P.M. blew very fresh with hard rain, which [*sic*] came in violent squalls, light-n'd a good deal and before Midnight wind veered to SSE and became more moderate. Pluv[iometer] 4.48 inches."

This storm is likely the same storm that passed quickly by Martinique on the evening of 13–14 August (Millás 1968). Millás estimates the storm was moving at up to 20 miles per hour. Assuming this is the same storm, it traveled approximately 1100 statute miles in 72 hours from near Martinique to a position abeam of Savanna-la-Mar from 10 P.M. to Midnight on 16 August.

[33] Thistlewood always abbreviated pluviometer as "Pluv" and recorded all his rainfall totals in fractional amounts (e.g., 210/100). I convert the fractional amounts into decimal equivalents and also append the measurement units (inches) throughout.

This would translate into an average speed of about 15.3 miles per hour. Given the short period of strongest winds at Savanna-la-Mar, this is consistent with a quick-moving storm. If it maintained a west-northwesterly course and speed, it most likely made landfall about midday on 18 August in the northernmost Yucatan Peninsula and then made a landfall somewhere in the northwest coastal areas of the Gulf of Mexico about 20 August.

15. THISTLEWOOD RECORD, MONSON COLLECTION 31/52, 17–18 OCTOBER 1766

Equivalent or related storm in Tannehill ([1938] 1952): October 22, Pensacola (Florida).

The remarkably active 1766 hurricane season brought a third tropical cyclone near Jamaica. The storm originated to the east of Jamaica and is first reported from Port Royal, Jamaica. A press item from Kingston, dated 18 October, states, "Yesterday the wind blew excessive hard, and occasioned a great swell of the sea, so that several vessels drove from their moorings and went on shore, and others lost their masts, yards, anchors, etc. Several small craft were wrecked. The storm began about nine o'clock in the morning and continued blowing when this paper went to press, so that we may expect to hear of considerable damage" *(Pennsylvania Gazette*, 27 November 1766).

The gale began about 12 hours later at Savanna-la-Mar. Thistlewood writes "At night fresh gales variable from East to ESE with hard squalls of wind and some rain."

> Saturday, 18th. Fresh gales E by S, before Noon variable to SE, cloudy, at times squalls of drizzling rain. In the evening, fresh breezes SSE with some hard squalls of wind and rain. At night, moderate.

This storm moved on a course probably to the north and northeast as a violent storm was felt at Pensacola on 22 and 23 October accompanied with a great sea. The storm produced NE to N to NW heavy gales, indicating a landfall to the east of Pensacola (HMS *Ferret* logbook, L/MAR/A/570F, Greenwich Maritime Museum).

16. THISTLEWOOD RECORD, MONSON COLLECTION 31/57, 23–24 MAY 1771

After the October 1766 gales, there were no tropical storm or greater force winds felt at Savanna-la-Mar until 1771. Several storms apparently passed near enough to produce suggestive patterns of wind direction fluctuations but no gales. Jamaica escaped the activity in the active 1768 hurricane season. Severe drought in Jamaica in 1769–1771 was in part a consequence of diminished hurricane activity in the Caribbean and the absence of storms near enough to Jamaica to be felt.

This storm appears to have formed west of Jamaica and produced hard squalls at Savanna-la-Mar from the afternoon of 23 May and "much the same at night," and squally weather of a less severe nature continued into 24 May. This storm appears to have deepened as it moved away from Jamaica and toward western Cuba. On 24 May, the sloop *Ann*, sailing from the Bay of Honduras for Bermuda, was wrecked on the south coast of Cuba in a hard gale of wind (*Pennsylvania Journal and Weekly Advertiser*, 8 August 1771). This tropical storm is believed to have continued to move north and contribute to the worst flood in Virginia history on 27 May 1771 (*Pennsylvania Journal and Weekly Advertiser*, 27 June 1771; Dennis Blanton, personal communication).

17. THISTLEWOOD RECORD, MONSON COLLECTION 31/58, 3 SEPTEMBER 1772

Equivalent or related storm in Millás (1968): Case 91, pp. 232–234. Case 92, pp. 235–238

Equivalent or related storm in Tannehill ([1938] 1952): (a) 1772, August 28, Puerto Rico, Jamaica; (b) August 31–September 4, Leeward Islands, Antigua, Virgin Islands, Puerto Rico, Jamaica. Louisiana.

This storm is probably a continuation of the great storm of 31 August at St. Croix and well documented in previous compilations of hurricanes. Thistlewood reported moderate breezes from the north all day on 2 September. By 3 September, the weather deteriorated further and the center of the hurricane appears to have passed to the north.

> Thursday 3ʳᵈ. Fresh gales varying almost continually between NW and SSW, cloudy, with frequent hard squalls of wind and rain & thunder remote. In the evening, the wind came about to about S By W from whence at night it blew fresh with at times hard squalls of wind and rain.

> Friday 4ᵗʰ. Fresh breezes SE, cloudy, some small showers of rain. P.M. somewhat squally with rain. Evening, and at night, moderate, cloudy.

Millás (1968) attributes the storm of 28 August that passed over the Leeward Islands to have been the one that continued westward near Jamaica and the Cayman Islands. But Thistlewood's weather diary gives no evidence to support this supposition. Instead, the second and more severe storm, which struck St. Croix on 31 August, was the one that passed to the north of Jamaica. This matches up with the account of Edward Long, quoted by Millás (1968), that the storm was felt with greatest force on the north coast of Jamaica and is supported by Thistlewood's account on the south shore of Jamaica.

Millás (1968), still quoting from Long (1774), shows that the hurricane probably passed to the north of Jamaica and on 3 September was probably between Cuba and the northwest tip of Jamaica as the storm center passed over a ship, with a thirty-minute calm. Long's description of the storm going "quite into the Bay of Honduras" does not fit with Millás's account of the storm then moving south into the Bay. Instead, it is probably based on other sources read by Long of ships that may have been driven southward by persistent northerly winds as the storm approached. The storm most likely moved to the west-northwest and passed over southwest Cuba and near the northern tip of the Yucatan Peninsula.

18. THISTLEWOOD RECORD, MONSON COLLECTION 31/64, 5 JUNE 1778

An early-season tropical storm was felt at Port Royal, Jamaica on 4 and 5 June (HMS *Southampton* ADM 51 914). At Savanna-la-Mar, the weather was "close and cloudy, at times drizzling rain" on the morning of the 4th. By night there were small showers of rain.

> Friday 5th. Close and cloudy. About 10 A.M. sprung up moderate gales SSE with frequent squalls of wind & rain. P.M. Fresh gales South & cloudy. At night hard squalls of wind with some rain, lightning & thunder at a distance.
>
> Saturday 6th. Fresh gales SSE, cloudy, some squalls of wind and rain, lightning & thunder remote. At night, light wind SSE & cloudy, lightning.

There are no other accounts of this early-season, previously undocumented, tropical storm.

19. THISTLEWOOD RECORD, MONSON COLLECTION 31/64, 16–17 SEPTEMBER 1778

This previously undocumented tropical storm affected all of Jamaica. HMS *Greenwich* (ADM 51 399) experienced the storm in port near Kingston on 15 and 16 September as it approached from the east. Thistlewood noted light airs with calms on the afternoon of the 14th as the hurricane approached from the east.

> Tuesday 15th. Moderate breezes NNE and at times cloudy. P.M. Light wind variable between NNE & East & most part cloudy. Thunder at a distance. In the evening, moderate breezes increasing till Midnight between NNE and ENE with lightning. Pluv[iometer] 2.82 inches.

> Wednesday 16th. From Midnight strong gales with hard squalls of wind & rain, variable from ENE to SE. Between 4 & 5 A.M. blew excessive hard with rain, an abundance of lightning some thunder just audible. About 7 A.M. some hard squalls of wind, rain, thunder remote and before noon somewhat abated. P.M. Fresh gales SE by S, cloudy, a few drops of rain. In the evening lightning and the wind fell entirely. Pluv[iometer] 2.26 inches.

The trajectory of this storm was probably between west and northwest, which would take it toward the northern Yucatan Peninsula. This storm produced some damage on Thistlewood's estate and was about as strong as the August 1766 storm, although the gusts were higher. Sustained winds were stronger only in the September 1751 and October 1764 hurricanes. This storm apparently gathered considerable strength, as the winds at Port Royal were not damaging.

20. THISTLEWOOD RECORD, MONSON COLLECTION 31/64, 5 OCTOBER 1778

Equivalent or related storm in Millás (1968): Case 109, p. 247.

Equivalent or related storm in Ludlum (1963): October 7–10, Gulf Coast

This hurricane approached Jamaica from the east, although no accounts of its effects east of Jamaica are yet available. The storm was felt at Port Royal, Jamaica on 4 and 5 October (HMS *Greenwich*, ADM 51 399). The second half of the 4th was rainy at Savanna-la-Mar.

> Monday 5th. From Midnight moderate breezes NE with rain. A little before 5 in the morning came on some hard squalls of wind & rain

from the ESE with thunder at a distance. Before 7 [A.M.] calm, but dark and cloudy with a mighty roaring of the sea & wind [round] to the Southward. About half past 8 A.M. sprang up fresh gales at SSE, cloudy, sometimes drizzling rain. P.M. The wind abated. At night moderate breezes, variable, cloudy, with some rain.

This storm apparently moved to the northwest and probably passed over western Cuba. This storm then made landfall about 9 October along the central Gulf coast of the present-day United States near Mobile, Alabama (Ludlum 1963).

21. THISTLEWOOD RECORD, MONSON COLLECTION 31/65, 25–26 MAY 1779

This is one of two tropical storms in the month of May to affect Savanna-la-Mar during the years 1751–1786. Like the other, this storm is previously undocumented.

> Tuesday 25th. Light airs ESE, cloudy, with moderate rain. [The rain] abated between 9 and 11, then began again. P.M. Fresh gales from the SE by S, with frequent hard squalls of wind & rain, which continued all night. Pluv[iometer] 0.99 inch.

> Wednesday 26th. Fresh gales SE with frequent hard squalls of wind and rain, thunder remote. About 8 A.M. strong gusts of wind from the SSE with hard rain. Towards noon abated somewhat. P.M. more moderate. At night, somewhat squally. Pluv[iometer] 2.66 inches.

Winds remained between SSE and SE the next several days. The storm probably moved due north toward central Cuba; however, no other accounts of this storm are available.

22. THISTLEWOOD RECORD, MONSON COLLECTION 31/66, 3 OCTOBER 1780

Equivalent or related storm in Millás (1968): Case 117, pp. 251–253.

Equivalent or related storm in Tannehill ([1938] 1952): 1780, October 3, Jamaica, Cuba.

The month of October 1780 produced three major hurricanes in the Caribbean region. The first storm probably originated from the region of the southern Leeward Islands and increased in strength as it approached Jamaica. HMS *Southampton* reported excessive strong squalls of wind and rain from midday the 2nd to midday the 3rd of October. At Savanna-la-Mar, the roar of the sea was heard on the 2nd, 24 hours before the hurricane set in. At this point, the storm was probably near the southeast coast of Jamaica and increasing in intensity.

Monday 2ⁿᵈ. Moderate breezes SSE and cloudy. About 11 A.M. a squall of wind & rain from SSE. P.M. Cloudy, with some small squalls of wind. The sea roars exceedingly & the sky looks very wild & dark and dismal in the south. Most part of the night moderate rain [wind] from the ENE. Pluv[iometer] 1.32 inches.

Tuesday 3ʳᵈ. Moderate breezes ENE, dark & cloudy with frequent showers of rain and a mighty roaring of the sea and a strong smell from [the sea]. Towards noon the wind shifted more easterly and kept still [sic] increasing. By 3 o'clock in the afternoon it blew violently from the ESE, but kept veering more Southerly. Before 4 it blew with indescribable force & fury and the continued hard rain, or rather streams of water, which came with it increased its power. It now roared dreadfully & the air was exceeding dark. About sunset or a little before, it came about to the South, or rather westward of the South, being then I think at its highest, raging with the utmost Violence & irresistible fury, in many places tearing up the surface of the Earth and making incredible ravage & ruin as it passed along. Soon after, it shifted almost in an instant, to the West, which continued without much abatement till near Midnight. During the hurricane, from between 3 & 4 in the afternoon, had dismal lightning in large pale sheets and in the night several shocks of earthquakes were felt, attended with a rumbling noise underground.

Great quantities of the spray of the sea was [sic] brought here by the wind. Pluv[iometer] I think fell 3.94, 4.06, 2.88 before 2 o'clock in the afternoon. Have allowed 22 inches in all. But as the water fell till past midnight in continued streams & so very quick succession, am persuaded far more than double that quantity fell. Suspect not less than 5 or 6 feet."

"Wednesday October 4ᵗʰ, 1780. Moderate breezes Westerly with frequent squalls of wind & rain. About 5 P.M. saw a rainbow in the East. Pluv[iometer] 5.58 inches.

Thursday 5ᵗʰ. Light wind, South, cloudy, some squalls of wind & rain. P.M. Cloudy. Pluv[iometer] 2.48 inches.

Several very interesting phenomena are mentioned in his weather diary. The center of the hurricane apparently passed only a few miles to the west of Breadnut Island, as there is no account of a calm. The tearing up of the Earth's surface could only be produced by tornadoes embedded within the spiral bands of the hurricane. The inundation of his property, and near total destruction of his great house and all his property, was caused by the storm surge, which swept through the town of Savanna-la-Mar and killed hundreds. Since the ground elevation of his second home at Breadnut Island is given as 15.5 feet above mean sea level, then we have a storm surge of at least 16 feet (and probably a bit higher). Storm surges of 15 feet are good evidence of a category 4 hurricane (on the Saffir-Simpson scale of hurricane intensity). Because wave action on top of the surge has to be considered, the actual damage effects are a combination of the surge and the waves, so a storm surge between 10 and 15 feet is more likely. Storm surges of this magnitude are typical of hurricanes producing sustained surface winds of 120 to 140 miles per hour. This suggests that the hurricane was a high-end category 3 or low-end category 4 storm.

At Montego Bay, about 25 miles north-northeast of Bread Nut Island, the weather in the first part of the afternoon of the 3rd was moderate and "much abated until between 3 and 4 o'clock in the evening, as to furnish no immediate indications of an approaching storm. About 4 o'clock, the wind seemed to be quite southerly, but increased (accompanied with incessant rain) to such an amazing degree, as about dark to threaten general ruin and destruction" (*Supplement to the Royal Gazette* [Kingston], 7 October 1780). The increase of wind was about four hours later than noticed by Thistlewood. This gives a speed of motion of the storm of about 6 or 7 miles per hour.

The hurricane was probably at its height between about 4 and 8 P.M. at Savanna-la-Mar. Another account from there indicated that " . . .about three o'clock the wind began to blow very hard from the south-east, accompanied with heavy rains, and by four had acquired such strength as to tear the trees up by the roots and strip the houses of their shingles. Between five and six, the sea began to rise, and continued for near an hour to swell to a most amazing height, overflowing the ill-fated town of Savanna-la-Mar and the low lands adjacent. From this time until eight o'clock, the force of the winds and the impetuosity of the waves, overthrew and demolished every house in that unfortunate place, and buried most of the inhabitants in its ruins. A little after eight, it began to abate, but nevertheless continued to blow very hard until midnight, when the wind veered round to the Westward . . ." (*Supplement to the Royal Gazette* [Kingston], 7 October 1780).

At Lucea, about 17 miles north by west of Bread Nut Island, the hurricane commenced with the wind from the north-east and blew hard to 6:00 P.M., a calm for half an hour followed, and was

then succeeded by the wind from south-west (Reid 1838). Winds
began to abate somewhat in the Savanna-la-Mar area at 8:00 P.M.
and at Lucea at 10:00 P.M. This indicates an approximate forward
motion of the hurricane at about 8 to 9 miles per hour. With the
hurricane moving at this rate, the hurricane center was about 4 to
4.5 miles in diameter. It appears to have made landfall only a short
distance west of Bread Nut Island (where no decrease in wind
speed was observed) and moving to the north-northeast. The hurricane center passed over Lucea and moved north-northeast
towards Cuba. At both Montego Bay and Breadnut Island, the
wind is considered to have diminished noticeably about midnight.
If this marked the outer edge of the hurricane force winds, then
(assuming a rate of motion of 8.5 miles per hour) hurricane force
winds extended some 52 miles to the south and southeast of its
center at midnight.

This was by far the greatest hurricane to strike Savanna-la-Mar
during Thistlewood's residence. In his other journal, he made the
following entry following the hurricane of 1 August 1781.

> "11ᵗʰ [old style] September 1751, Violence or Force, NOT
>   Velocity, say 6
> 3ʳᵈ October 1780, say 10
> 1ˢᵗ August 1781, say 4 1/2
> The greatest I ever saw at home [Lincolnshire, England],
>   about 3." (Hall 1989)[34]

It is apparent from Table 21 that the wind speed, as derived from
his wind force terms, was clearly a cut above the other great hurricanes of 1751 and 1781 and was the most powerful Jamaican
hurricane of the eighteenth century. The center of the storm
passed over Lucea Bay (Millás 1968), which is slightly west of due
north from Breadnut Island. This storm then moved over eastern
Cuba and the central Bahamas. It was felt by HMS *Hector* between
latitude 28° and 29° N and longitude 72° to 73° W , to the northeast of the Bahamas (Reid 1838).

---

[34] With respect to the 3 October 1780 hurricane, it is assumed that winds of about
130 miles per hour (mph) were felt. Using the equation $P = KV^2$, to measure the
pressure of the wind in pounds per square inch (ppsi) (P is the pressure exerted
by the force of the wind and V is the wind speed in mph) and a coefficient value
of K of 0.0038 gives a value of 64.2 ppsi. Multiplying the relative force values
given by Thistlewood of 0.65 for the 1751 hurricane would give a value of 38.5
ppsi, or a wind speed of almost 101 mph. For the 1781 hurricane, the corresponding figures are 28.9 ppsi and 87 mph. For the maximum wind at Lincolnshire,
U.K. the figures are 19.3 ppsi and 71 mph. Note that none of the other hurricanes
came so close to Savanna-la-Mar as the 1780 hurricane, so their actual maximum
wind speeds are not known, only their relative strength at Savanna-la-Mar.

23. THISTLEWOOD RECORD, MONSON COLLECTION 31/66, 16–17 OCTOBER 1780

Equivalent or related storm in Millás (1968): Case 119, pp. 260–262.

Equivalent or related storm in Tannehill ([1938] 1952): 1780, October 16,17, Cuba. Solano's Storm.

The Great Hurricane of October 10–16 affected the Leeward Islands and Bermuda but left Jamaica untouched. The third major hurricane affected Jamaica, but fortunately its most severe impact was felt to the northwest of Jamaica. Millás (1968) wrote that the storm formed over the southeastern Gulf of Mexico. Thistlewood's record shows that the storm originated to the east of Jamaica, probably coming from an incipient disturbance that formed to the west of the slow-moving major hurricane over the Lesser Antilles, and developed rapidly once the Great Hurricane had exited the immediate region by the 15th. This, in turn, is supported by the logbook of HMS *Pelican* (ADM 51 710), south of the eastern tip of Jamaica from 15–17 October, where west and southwest winds of up to hard gales were observed.

> Sunday 15th. Light winds variable and at times cloudy, thunder at a distance. Soon after Noon a small shower of rain, then moderate breezes North at times cloudy. Pluv[iometer] 0.02 inch.

> Monday 16th. Light winds North & cloudy. Afterwards variable to NW and West and cloudy. At night frequent hard squalls of wind with some rain, thunder and lightning from the WNW. Pluv[iometer] 0.93 inches.

> Tuesday Seventeenth. In the morning moderate breezes NW and cloudy. Middle of the day fresh breezes West, somewhat cloudy & hazy, looks exceeding wild & ugly and the sea roared much this morning. About 2 P.M. some squalls of wind and rain from the WSW and the sea roars very loud in the SW also. In the evening, moderate breezes WSW & cloudy, looks dismally about the foot of Bluefields. Pluv[iometer] 0.70 inch.

> Wednesday Eighteenth. Fresh breezes SSW and a good deal cloudy. At times squalls of wind and rain. Pluv[iometer] 0.24 inch.

This storm continued on to the northwest, after passing north of Jamaica, and destroyed the Spanish fleet sailing toward Florida on 20–21 October (Millás 1968).

24. THISTLEWOOD RECORD, MONSON COLLECTION 31/67, 1–2 AUGUST 1781

Equivalent or related storm in Millás (1968): Case 121, pp. 262–263.

Equivalent or related storm in Tannehill ([1938] 1952): 1781, August 1, Jamaica. Passed near Kingston.

The damage done by the hurricane of October 1780 in Jamaica was so severe that the British government sent an emergency supply of specie aboard HMS *Ulysses* to be delivered to Savanna-la-Mar for the relief of its citizens. The *Ulysses* reached Savanna-la-Mar at the end of July 1781 and was caught up in a hurricane exceeded in strength only by the 1751 and 1780 hurricanes. Fortunately, the ship did not founder and was able to deliver its monies to the distressed inhabitants (*The Royal Gazette* [Kingston], 25 August 1781).

This hurricane came from the east and was felt throughout the island of Jamaica. The telltale northerly winds began at Savanna-la-Mar on 31 July and the sea roared on the reef offshore of Savanna-la-Mar that evening.

> Wednesday, August 1, 1781. Light wind NE by E and most part fair. P.M. Cloudy, moderate rain, thunder at a distance. After about 8 o'Clock at night, heard no more thunder but soon had some uncommon hard gusts of wind from the N by E and presently strong gales with hard squalls of wind & rain from N by E, which kept increasing with an abundance of lightning. In the evening the reef roared very loud. Pluv[iometer] 8.27 inches.

> Thursday, 2ⁿᵈ. From Midnight, strong gales, dark & cloudy with hard squalls of wind & rain from about NNE, which still kept increasing & shifting more Easterly and from about 2 o'Clock in the morning till day break, the wind roared like thunder and blew most excessive hard, veering to East, SE and South & even to the Westward of the South, attended with a great deal of lightning. Seemed to be at its height between 4 & 5 o'Clock, after sunrise abated somewhat but continued to blow very strong from the S by E with frequent squalls of rain. P.M. Fresh breezes and cloudy. Evening moderate. Pluv[iometer] 0.53 inch.

The trajectory of this storm probably took it toward the Yucatan Channel and perhaps an eventual landfall on the Texas Gulf Coast.

25. THISTLEWOOD RECORD, MONSON COLLECTION 31/67, 6–7 SEPTEMBER 1781

Equivalent or related storm in Millás (1968): Case 122, p. 264.

Equivalent or related storm in Tannehill ([1938] 1952): 1781, September 5, Santo Domingo.

Two tropical cyclones produced sustained tropical storm- or hurricane-force winds at Savanna-la-Mar in 1781 and two others produced tropical storm force gusts and squalls. The second storm to affect Thistlewood is the same storm that Millás mentions that passed by Santo Domingo on 5 September. The storm was felt at Port Royal, Jamaica by HMS *Southampton* (ADM 51 914) on 5 and 6 September. At Savanna-la-Mar, signs of the storm were evident on the 5th but the effects of the storm were not felt there until the 6th.

> Thursday 6th. Calm, dark & cloudy. About 8 A.M. sprang up moderate breezes variable between SSW & NW. Cloudy with some small showers of rain. P.M. Fresh breezes S by E and cloudy. At night, strong gales from the same quarter, with frequent hard squalls of wind and some few drops of rain. It would sometime be perfectly calm for 5 or 6 minutes, then return with redoubled fury. Note: about $1/2$ past 5 in the evening, a mock sun to the Southward of the sun. The wind roared a good deal in the night. Pluv[iometer] 0.07 inch.

> Friday 7th. Fresh breezes S by E, cloudy, with frequent hard squalls of wind & some rain. P.M. Some very hard squalls of wind & rain, thunder remote. About 5 in the evening, almost calm, which continued till about 9 P.M. when had some hard gusts of wind from the SSE attended with a mighty roaring like thunder at a distance. Cloudy & a good deal of lightning. Pluv[iometer] 0.37 inch.

This hurricane approached from the east and passed to the north of Jamaica. It apparently strengthened in intensity after passing by Jamaica, because the weather remained stormy for some time after the center had passed by to the north. There is no record of a storm at Havana, Cuba so the storm may have stayed to the southwest of Havana.

26. THISTLEWOOD RECORD, MONSON COLLECTION 31/67, 8 OCTOBER 1781

The third storm of the 1781 hurricane season felt at Savanna-la-Mar is a previously undocumented storm. The storm appears to have formed in the Western Caribbean and passed west of Jamaica. It appears to have moved to the northeast and may well be the same storm that affected Bermuda on 13 and 14 October (HMS *Hornet*, ADM 51 459).

> Monday 8ᵗʰ. Light winds SSE and at times cloudy. P.M. Cloudy, a shower of rain, thunder at a distance. Between 6 and 7 in the evening the sea roared exceedingly and about 7 in the evening sprang up all at once, strong gales with hard gusts of wind, attended with a few drops of rain from the SSE which continued for about 2 hours then abated and was a fine night thereafter. Pluv[iometer] 0.10 inch.

27. THISTLEWOOD RECORD, MONSON COLLECTION 31/67, 2–3 NOVEMBER 1781

This previously unrecorded storm brushed by Savanna-la-Mar on the night of 2–3 November. The storm developed to the west of Jamaica. On 1 November, light variable winds from south to southwest to west were observed. On 2 November, the day was cloudy with calm or very light variable winds. That evening, a light wind from the South was established with "frequent strong puffs of wind that were over in half a minute, & roared as they went along, then almost calm for 10 or 12 minutes, when another puff came, and so continued all night."

The tropical storm probably meandered to the west of Jamaica. On 4 November, light winds and some squalls were noted from SE by E and the sky had "a very ugly appearance" during the day. On 5 November, "the sea roared a good deal" and winds in the afternoon were moderate from the South. After this date, the weather remained somewhat disturbed with light to moderate southerly breezes. The tropical storm probably degenerated over the northwest Caribbean.

28. THISTLEWOOD RECORD, MONSON COLLECTION 31/68, 30 JUNE 1782

This previously undocumented storm appears to have formed to the southwest of Jamaica, but may have been dissipating by the time it approached Savanna-la-Mar.

> Sunday 30ᵗʰ. Moderate breezes, variable, and at times cloudy. P.M. Cloudy, thunder at a distance. The sea roared very much all day and between 9 and 10 o'Clock at night came on a very hard squall of

wind, with a mighty roaring from the SE which held about an hour, then abated, the sky somewhat cloudy, with an abundance of lightning & thunder far off.

Monday July 1ˢᵗ 1782. Moderate breezes variable all round [the compass] and at times cloudy, lightning thunder remote. P.M. Hazy. At night, lightning.

29. THISTLEWOOD RECORD, MONSON COLLECTION 31/69, 4–5 OCTOBER 1783

Equivalent or related storm in Ludlum (1963): October 7–8, Entire (US) Coast

This tropical cyclone is apparently the same storm as documented by Ludlum (1963) as running up the entire eastern coast of the United States on 7 and 8 October. If so, the storm originated in the northwest Caribbean and pursued a north-northeast to northeast course from its birth region.

Saturday 4ᵗʰ. Morning & Evening light winds, variable and at times cloudy. Middle of the day fresh breezes South & cloudy, thunder at a distance. At night, fresh gales variable between SE and SSW with hard squalls of wind & rain, lightning & thunder remote. Pluv[iometer] 0.22 inch.

Sunday 5ᵗʰ. Fresh gales SSE, cloudy, thunder far off. At night, hard squalls of wind with some rain from the SE. Pluv[iometer] 0.28 inch.

The disturbed weather, with squalls of wind and rain and southerly winds, continued through 9 October.

30. THISTLEWOOD RECORD, MONSON COLLECTION 31/70, 30–31 JULY 1784

Equivalent or related storm in Millás (1968): Case 126, p. 266.

Equivalent or related storm in Tannehill ([1938] 1952): 1784, July 30. Jamaica, Santo Domingo.

The hurricane of July 1784 was the third hurricane to have a major economic impact on the island of Jamaica in the 1780s. It was a severe hurricane that wrecked many ships at Port Royal, Jamaica on the evening of 30 July (Millás 1968). Thistlewood shared in suffering damage from the hurricane, but destruction was much less than at Port Royal.

Friday 30ᵗʰ. Moderate breezes NNE and at times cloudy. Evening cloudy, thunder at a distance and a great deal of lightning. Between 8 and 9 o'Clock at night sprang up fresh gales between North & NNE with frequent hard squalls of wind & rain with thunder and lightning. The lightning continued but soon after 9 o'Clock such a roaring of the wind could not be certain whether it thundered or not. Pluv[iometer] 9.05 inches. As soon as light this morning, there were ugly ragged clouds, which foreboded an approaching North, and all the day very suspicious weather.

Saturday 31ˢᵗ. Strong gales with hard squalls of wind & rain from about the NNE but by about 3 A.M. was come about to the East, soon after varied to the SE and by day back to SSE, all along attended with frequent hard squalls of wind & rain and an abundance of lightning. Continued all day. From about 3 A.M. till day break blew excessive hard and the rain poured down. Continued fresh gales between SE and SSE, cloudy. Evening moderate. At night, lightning, thunder far off.

At Montego Bay the storm began between 9 and 10 P.M. on the 30ᵗʰ and blew hard from the NE, but later shifted to the SE. Violent squalls from the SE continued until 3 A.M. on the 31st. There was no significant damage reported. At Kingston, the height of the storm raged near three hours from about 8:30 P.M. "until past eleven, when it moderated" according to the *Kingston Post* of 31 July (*Supplement to the Cornwall Chronicle*, 7 August 1784). The center of the hurricane apparently passed very near Kingston as a later press item from Kingston, dated 4 August, indicates. "The tempest, on Friday night, was the fiercest, for about an hour, that has been known in the memory of the oldest man living in this island; and, had its duration been in proportion to its violence, the whole town, in all probability, would have been a heap of ruins. The raging of the sea, the warring of the winds, which shifted to every point of the compass in a few minutes . . . " (*Supplement to the Cornwall Chronicle*, 14 August 1784).

The strongest winds at Savanna-la-Mar also lasted a similar length of time (about three hours) but were about 6.5 hours later. The storm began about one hour later at Montego Bay and ended at about the same time. This suggests a slight inclination of the storm track slightly to the north of west. This westward-moving hurricane was then moving at a speed of almost 15 miles per hour. The hurricane probably made landfall in the Yucatan Peninsula.

At just about the time the strongest winds affected Kingston, gale force winds began at Savanna-la-Mar. Assuming a movement of 15 miles per hour, the hurricane center was at some distance abeam of shore about 22.5 miles east of Kingston at around

8:30 P.M. This allows a calculation of the approximate diameter of sustained tropical storm-force winds to extend to the northwest of the center by about 110 statute miles. It seems that the hurricane center passed over the southern part of Jamaica and passed back out to sea again south of Bluefields.

31. THISTLEWOOD RECORD, MONSON COLLECTION 31/71, 28 AUGUST 1785

Equivalent or related storm in Millás (1968): Case 131, pp. 268–269. Case 132, pp. 269–270.

Equivalent or related storm in Tannehill ([1938] 1952): (a) 1785, August 25, Guadeloupe, St. Kitts; (b) 1785, August 27, Jamaica. Possibly same as proceeding.

This well-documented storm produced heavy damage in the Leeward Islands on 25 August and two days later in eastern Jamaica (Millás 1968). This was the fourth violent storm at Savanna-la-Mar since October 1780.

> Saturday 27th. In the morning light winds NNW & cloudy. Afterwards fresh breezes North & cloudy. P.M. Strong gales North, some squalls of wind, with a few drops of rain and thunder at a distance. At night, very strong gales with frequent hard gusts of wind & rain from North and frequent lightning. Pluv[iometer] 3.78 inches.

> Sunday 28th. Strong gales of wind with hard gusts of wind & rain from the North & at times lightning. About 2 o'Clock in the morning very violent and by 3 [o'Clock] varied to the SE from whence blew very strong with frequent hard squalls of wind and rain. About 11 A.M. rather abated. P.M. fresh gales SE & cloudy with some small showers of rain. Pluv[iometer] 3.26 inches.

This storm apparently continued to move to the west or west-northwest, as its further ravages remain unknown from English-language sources.

32. THISTLEWOOD RECORD, MONSON COLLECTION 31/71, 25 OCTOBER 1785

This tropical storm came from the east of Jamaica and gained strength as it passed Jamaica. Thistlewood reported fresh gales but with hard squalls of wind during the afternoon and evening of 25 October. Winds that morning were from the north-northeast, but switched to the southeast by 9 A.M. and to the south from noon, where it remained fixed the remainder of the day. On the 26th, fresh gales from S by W prevailed all day with some squalls of wind and rain. On the 27th, the weather moderated and moderate breezes from SW prevailed.

The only other record of this storm comes from the *Cornwall Chronicle*. "Tuesday night [25ᵗʰ] there came on a heavy swell of the sea, at N.W. which raged with great violence all the next day, and did considerable damage to the wharfs [*sic*] and houses on the beach . . .. The shipping rode it out safe, the wind happily blowing from the Southward . . ." (*Supplement to the Cornwall Chronicle*, 29 October 1785). In St. Elizabeth Parish (midway between Savanna-la-Mar and Kingston on the southern coast) incessant rains began on 24 October and severe flooding followed (*Supplement to the Cornwall Chronicle*, 19 November 1785).

This storm was probably a tropical storm that passed to the south of Jamaica on a course to the west-northwest, and by the 27th was probably moving to the northwest. At this late date in the hurricane season, the storm was likely soon pulled more to the north and then to the northeast over Cuba and then Florida or the Bahamas.

33. THISTLEWOOD RECORD, MONSON COLLECTION 31/72, 5 JUNE 1786

Hurricanes in the months of May and June are relatively infrequent, and since most compilations of Atlantic tropical cyclones are of hurricane intensity, we have few records of early season activity. For this reason, all of the early-season storms observed by Thistlewood are documented here for the first time. The absence of surviving local newspapers has also contributed to the lack of documented early-season storms. One surviving paper from Jamaica does provide additional data for an early-season tropical cyclone that affected Savanna-la-Mar on 5 June. Thistlewood wrote:

> Monday 5ᵗʰ. Fresh gales variable between SSE and S by W, dark and cloudy with frequent hard squalls of wind & rain. At night frequent hard showers with an abundance of thunder and lightning. Between sunrise and noon fell 0.95 inch; between Noon & sunset 2.08 inches and to sunrise of Tuesday morning 3.45 inch. [Grand total] 6.48 inches.

> Tuesday 6ᵗʰ. Moderate breezes variable between S by E and S by W, dark and cloudy with almost continual hard rain attended with thunder and lightning. About 3 P.M. [rain] broke up. Evening,

calm, cloudy, thunder remote. At night, lightning. Pluv[iometer] 3.16 inches.

This tropical storm came after lighter drought-breaking rains fell on 3 and 4 June. The ship *Vigilant* arrived at Bluefields on 4 June and the captain traveled in the long boat to Savanna-la-Mar to clear the ship. On 5 June, he was detained at Savanna-la-Mar by [the arising of] "such a violent breeze, and extreme weather, that rendered it impossible for him to return to Bluefields" [until the evening of the 7th of June] (*Supplement to the Cornwall Chronicle*, 24 June 1786). During his absence his ship had been blown out to sea. Severe flooding accompanied the storm in eastern Jamaica on 6 June.

"We hear from Clarendon, that on Tuesday the 6th instant, the Rio Mino, which rises in the mountains of that parish, and which was greatly swelled by the heavy rains that had continued for several days before, suddenly overflowed its banks and with irresistible violence swept away every building on Windsor Estate, the property of Mr. Dawkins, so that not the least vestige of a habitation remains, together with a considerable quantity of livestock. The river now runs through the midst of the estate and the damage occasioned by the inundation is computed at an enormous sum. Another sugar estate in the neighborhood, it seems, lost 30 hogshead of sugar by the same floods." (*Supplement to the Cornwall Chronicle*, 24 June 1786)

As with most early June tropical storms the center formed in the western Caribbean and remained to the west of Jamaica. There is another account in the *Cornwall Chronicle* of 19 August 1786 "that there had been several shocks of an earthquake felt in the eastern part of Hispanolia in June last, attended with heavy rains and violent gales of wind." If so, this must have been another storm as there is utterly no evidence in Thistlewood's record that would suggest the early June storm came from the east of Jamaica.

34. THISTLEWOOD RECORD, MONSON COLLECTION 31/72, 20 OCTOBER 1786

Equivalent or related storm in Millás (1968): Case 140, pp. 273–274.

Equivalent or related storm in Tannehill ([1938] 1952): 1786, October 20. Jamaica.

The last of a series of five major hurricanes to strike Jamaica in the 1780s struck on 20 October. Thistlewood had less than two months to live at this time and was ill. Nonetheless, his account of the storm is as complete and reliable as any before.

> Thursday 19th. Light winds North & at times cloudy. P.M. Moderate [breezes] & a good deal cloudy. At night, some hard squalls of wind and rain. Very suspicious [weather]. Pluv[iometer] 0.32 inch.
>
> Friday 20th. About sunrise moderate breezes North, dark & cloudy. Then came on hard gales with frequent squalls of wind between North & NNE, which kept increasing till about 11 A.M. when it blew excessive hard and came about by degrees more Easterly and by half past 12 wind ESE and soon after SE where it raged with great violence till about 5 P.M. but was at its height between 1 & 3. In the evening came to the South & SSW and rather abated but continued all night hard gales with frequent squalls of wind & rain.
>
> Pluv[iometer] between sunrise & sunset 2.40 inches. Between sunset & sunrise Saturday morning 1.85 inches = 4.25 inches. This is by far the worst since the great hurricane in 1780, tho' not to compare with that.
>
> Saturday 21st. Fresh gales of wind with frequent squalls from the South, lightning and thunder at a distance and continued at night. Pluv[iometer] all the day 46/100 and night 140=1.86 [inches].

The *Cornwall Chronicle* agreed with Thistlewood's assessment of the damage in his part of Jamaica. The newspaper states that the squalls began between 11 P.M. and midnight on the 19th, blowing from the east. By daylight on Friday, "the wind shifted to the southward from which quarter it blew with great violence all Friday, and rained incessantly . . . Our accounts from Hanover and Westmoreland [where Thistlewood resided] are of a more alarming nature. Many of the estates in the leeward part of each parish are equally in as bad a situation as after the hurricane of October 1780 . . . At Savanna-la-Mar all the small craft are ashore; a ship (the only one there) rode it out. The town has suffered a good deal. Mr. Antrobus's house and some others are blown down, and a great many damaged in the roofs. A letter from Westmoreland mentions, 'The appearance every where demonstrates the superior vehemence of this gust over all we have experienced since 1780. The trees, stripped of their leaves, exhibit an appearance as if fire had devoured their verdure. The shores are covered with the dead bodies of duck, teal, and other aquatic birds that have been driven with irresistible impetuosity against the trunks of mangroves and dashed to pieces'" (*Supplement to the Cornwall Chronicle*, 28 October 1786).

REFERENCES

Allan, R.J., 2000. ENSO and climatic variability in the past 150 years. In *El Niño and the Southern Oscillation*, eds. H. F. Diaz and V. Markgraf. Cambridge: Cambridge University Press, 3–55.

Allan, R.J., and R.D. D'Arrigo. 1999. Persistent ENSO sequences: How unusual was the 1990–1995 El Niño? *Holocene* 9:101–118.

Baron, W. 1991. A reconstruction of the Holyoke temperature record for Salem, Massachusetts 1754–1829. *Wurzburger Geographische Arbeiten* 80:183–198.

Berry, F.A., Jr., E. Bollay, and N.R. Beers, eds. 1945. *Handbook of meteorology*. New York: McGraw-Hill, 884–901.

Bottomley, M., C.K. Folland, J. Hsuing, R.E. Newell, and D.E. Parker. 1990. *Global ocean surface temperature atlas*. Bracknell: Meteorological Office and Massachusetts Institute of Technology.

Bove, M.C., J.B. Elsner, C.W. Landsea, X. Niu, and J.J. O'Brien. 1998. Effect of El Niño on U.S. landfalling hurricanes, revisited. *Bulletin of the American Meteorological Society*, 79:2477–2482.

Bradley, R.S., and P.D. Jones, (eds.). 1992. *Climate since A.D. 1500*. London: Routledge.

Byam, F. 1755. Rainfall record for Antigua 1751–4. *Philosophical Transactions of the Royal Society* 10:628.

Chang, P., L. Ji, and H. Li. 1997. A decadal climate variation in the tropical Atlantic Ocean from thermodynamic air-sea interactions. *Nature* 385:516–518.

Chen, A., A. Roy, J. McTavish, M. Taylor,. and L. Marx. 1997. Using SST anomalies to predict flood and drought conditions in the Caribbean. Institute of Global Environment and Society, *Report No. 49*, Center for Ocean-Land-Atmosphere Studies, Calverton, MD.

Chenoweth, M.1993. Non-standard thermometer exposures at U.S. cooperative weather stations during the late nineteenth century. *Journal of Climate* 6:1787–1797.

—————. 1998. The 19th Century climate of the Bahamas and a comparison with 20th Century averages. *Climatic Change* 40:577–603.

—————. 1999. Historical marine data in U.K. and American archives. In: *Proceedings of the International Workshop on Digitization and Preparation of Historical Marine Data and Metadata*, eds. H.F. Diaz and S.D. Woodruff. World Meteorological Organization/Technical Document No. 957, 57–60.

—————. 2000. A new methodology for homogenization of 19th century marine air temperature data. *Journal of Geophysical Research* 105:29145–29154.

—————. 2001. Two major volcanic cooling episodes derived from global marine air temperature, AD 1807–1827. *Geophysical Research Letters* 28:2963–2966.

Cry, G.W. 1967. Effects of tropical cyclone rainfall on the distribution of precipitation over the eastern and southern United States. *ESSA Professional Paper 1*, U.S. Department of Commerce, Environmental Sciences Service Administration, Washington, DC.

Dherent, C., and G. Petit-Renaud. 1994. Using archival resources for climate history research. *World Meteorological Organization, UNESCO, IHP-IV, SC-94/WS.6*, Paris.

Diaz, H.F. 1991. Some characteristics of wet and dry regimes in the contiguous United States: implications for climate change detection efforts. In *Greenhouse-Gas-Induced Climatic Change*, M.E. Schlesinger, ed. Elsevier, Amsterdam: 269–296.

Dunn, R.S. 1972. *Sugar and slaves. The rise of the planter class in the English West Indies, 1624–1713*. Chapel Hill, University of North Carolina Press.

Easterling, D.R., T.R. Karl, E. H. Mason, P.Y. Hughes, D. P. Bowman, R. C. Daniels, and T. A. Boden, eds. 1996. United States Historical Climatology Network (USHCN) monthly temperature and precipitation data. *ORNL/CDIAC-87, NDP-019/R3*. Carbon Dioxide Information Analysis Center, Oak Ridge National Laboratory, Oak Ridge, TN.

Eischeid, J.K., H. F. Diaz, R.S. Bradley, and P.D. Jones. 1991. A comprehensive precipitation data set for global land areas. Carbon Dioxide Information Analysis Center, *TR-051*, Oak Ridge National Laboratory, Oak Ridge, TN.

Elsner, J.B., and A.B. Kara. 1999. *Hurricanes of the North Atlantic: Climate and society*. Oxford: Oxford University Press.

Elsner, J.B., G.S. Lehmiller, and T.B. Kimberlain. 1996. Objective classification of Atlantic hurricanes. *Journal of Climate* 9:2880–2889.

Enfield, D.B., and A.M. Mestas-Nuñez. 2000. Global modes of ENSO and non-ENSO sea surface temperature variability and their associations with climate. In *El Niño and the Southern Oscillation*, eds. H. F. Diaz and V. Markgraf. Cambridge: Cambridge University Press, 89–112.

Enfield, D.B., and D.A. Mayer, 1997. Tropical Atlantic SST variability and its relation to El Niño-Southern Oscillation. *Journal of Geophysical Research* 102:929–945.

Fernández-Partágas, J. and H.F. Diaz, 1995a. *A reconstruction of historical tropical cyclone frequency in the Atlantic from documentary and other historical sources, 1851 to 1880. Part I: 1851–1870*. Climate Diagnostics Center, Environmental Research Laboratories, National Oceanic and Atmospheric Administration, Boulder, CO..

—————, 1995b. *A reconstruction of historical tropical cyclone frequency in the Atlantic from documentary and other historical sources, 1851 to 1880. Part II: 1871–1880*. Climate Diagnostics Center, Environmental Research Laboratories, National Oceanic and Atmospheric Administration, Boulder, CO.

—————, 1996a. *A reconstruction of historical tropical cyclone frequency in the Atlantic from documentary and other historical sources, Part III: 1881–1890*. Climate Diagnostics Center, Environmental Research Laboratories, National Oceanic and Atmospheric Administration, Boulder, CO.

—————, 1996b. *A reconstruction of historical tropical cyclone frequency in the Atlantic from documentary and other historical sources. Part IV: 1891–1900*. Climate Diagnostics Center, Environmental Research Laboratories, National Oceanic and Atmospheric Administration, Boulder, CO.

—————, 1997. *A reconstruction of historical tropical cyclone frequency in the Atlantic from documentary and other historical sources. Part V: 1901–1908*. Climate Diagnostics Center, Environmental Research Laboratories, National Oceanic and Atmospheric Administration, Boulder, CO.

—————, 1999. *A reconstruction of historical tropical cyclone frequency in the Atlantic from documentary and other historical sources. Part VI: 1909–1910*. Climate Diagnostics Center, Environmental Research Laboratories, National Oceanic and Atmospheric Administration, Boulder, CO.

Fleming, J.R. 1990. *Meteorology in America, 1800–1870*. Baltimore: The Johns Hopkins University Press.

Folland, C.K., T.N. Palmer, and D.E. Parker. 1986. Sahel rainfall and worldwide sea temperatures, 1901–1985. *Nature* 320:602–607.

Glaser, R., S. Militzer, and R. Walsh. 1991. Weather and climate at Madras, in the years 1732–1737 based upon an analysis of the weather diary of the German missionary Geisler. *Wurzburger Geographische Arbeiten* 80:45–86.

Goldenberg, S.B., and L.J. Shapiro. 1996. Physical mechanisms for the association of El Niño and West African rainfall with Atlantic major hurricane activity. *Journal of Climate* 9:1169–1187.

—————, C.W. Landsea, A.M. Mestas-Nunez, and W.M. Gray. 2001. The recent increase in Atlantic hurricane activity: causes and implications. *Science* 293:474–479.

Gray, W.M. 1984. Atlantic seasonal hurricane frequency. Part I: El Niño and 30 mb quasi-biennial oscillation influences. *Monthly Weather Review* 112:1649–1668.

—————. 1990. Strong association between West African rainfall and US landfall of intense hurricanes. *Science* 249:1251–1256.

Guttman, N.B. 1989. Statistical descriptors of climate. *Bulletin of the American Meteorological Society* 70:602–607.

Hadeen, K., and R. Davis. 1990. Metadata requirements and standards for global baseline data sets. *Meeting on Global Baseline Data Sets, 22–26 January 1990*, Asheville, NC.

Hall, D., 1989. *In miserable slavery. Thomas Thistlewood in Jamaica, 1750–86*. Warwick University Caribbean Studies, New York: The Macmillan Press Ltd, 1989.

Halpert, M.S. and C.F. Ropelewski. 1992. Surface temperature patterns associated with the Southern Oscillation. *Journal of Climate* 5:577–593.

Hansen, D.V., and H.F. Bezdek. 1996. On the nature of decadal anomalies in North Atlantic sea surface temperature. *Journal of Geophysical Research* 101:8749–8758.

Harrison, D.E., and N.K Larkin. 1998. El Niño-Southern Oscillation sea surface temperature and wind anomalies, 1946–1993. *Reviews of Geophysics* 36,3:353–399.

Hoerling, M.P., and A. Kumar. 2000. Understanding and predicting extratropical teleconnections related to ENSO. In *El Niño and the Southern Oscillation*, eds. H.F. Diaz and V.Markgraf. Cambridge: Cambridge University Press, 58–88.

Hurrell, J.W. 1995. Decadal trends in the North Atlantic Oscillation: Regional temperatures and precipitation. *Science* 269:676–679.

Hydrographic Office of the Navy. 1993. *West Indies Pilot, Vol. 1*. Taunton, Somerset, England: Hydrographic Office of the Navy, Hydrographic Office, MOD Taunton.

Jamaican Meteorological Service. 1973. *The Climate of Jamaica*. Kingston: Climatological Branch of the Meteorological Service, First Edition.

Jarvinen, B. R., C. J. Neumann, and M. A. S. Davis. 1984. A tropical cyclone data tape for the North Atlantic basin, 1886–1983: Contents, limitations, and uses. NOAA Tech. Memo. NWS NHC 22, Coral Gables, Florida, 21 pp.

Jones, P.D., M. New, D.E. Parker, S. Martin, and I.G. Rigor. 1999. Surface air temperature and its changes over the past 150 years. *Reviews of Geophysics* 37:173–199.

Kalnay, E., M. Kanamitsu, R. Kistler, W. Collins, D. Deaven, L. Gandin, M. Iredell, S. Saha, G. White, J. Woolen, Y. Zhu, M. Chelliah, W. Ebisuzaki, W. Higgins, J. Janowiak, K.C. Mo, C. Ropelewski, J. Wang, A. Leetma, R. Reynolds, R. Jenne, and D. Joseph. 1996. The NCEP/NCAR 40-year reanalysis project. *Bulletin of the American Meteorological Society* 77:437–471.

Karl, T.R., and C.N. Williams. 1987. An approach to adjusting climatological time series for discontinuities. *Journal of Climate and Applied Meteorology* 26:1744–1763.

Karl, T. R., C.N. Williams, P. J. Young, and W.M. Wendland. 1986. A model to estimate the time of observation bias associated with monthly mean maximum, minimum and mean temperatures for the United States. *Journal of Climate and Applied Meteorology* 25:145–160.

Kinsman, B. 1969. Who put the wind speed in Admiral Beaufort's force scale? *Oceans 2*, 2:18–25.

Knaff, J.A. 1997. Implications of summertime sea level pressure anomalies in the tropical Atlantic region. *Monthly Weather Review* 125:789–804.

Lamb, H. 1977. *Climate, Present, Past and Future, Vol. 2*. New York: Methuen & Co.

Lamb, H., and K. Frydendahl. 1991. *Historic Storms of the North Sea, British Isles and Northwest Europe*. Cambridge: Cambridge University Press.

Landsberg, H.E., C.S. Yu, and L. Huang. 1968. Preliminary reconstruction of a long time series of climatic data for the eastern United States. *Tech. Note B-571*. College Park: University of Maryland, Institute of Fluid Dynamics.

Landsea, C.W., W.M. Gray, P.W. Mielke, Jr., and K.J. Berry. 1992. Long-term variations of Western Sahelian monsoon and rainfall and intense U.S. landfalling hurricanes. *Journal of Climate* 5:1528–1534.

Landsea, C.W., C. Anderson, G. Clark, J. Fernández-Partágas, P. Hungerford, C. Neumann, and M. Zimmer, 1998. The Atlantic hurricane database re-analysis project. Boston: American Meteorological Society, 23rd Annual Conference on Hurricanes and Tropical Meteorology, 394–397.

Landsea, C.W., R.A. Pielke, Jr., A.M. Mestas-Nuñez, and J.A. Knaff. 1999. Atlantic basin hurricanes: Indices of climatic change. *Climatic Change* 42:89–129.

Le Barbe, L., T. Lebel, and D. Tapsoba. 2002. Rainfall variability in West Africa during the years 1950–90. *Journal of Climate* 15:187–202.

Long, Edward. 1774. *The History of Jamaica, Vol. III.* (Arno Press 1972 Reprint Edition of 1774 Original).

Ludlum, D.M. 1963. *Early American Hurricanes, 1492–1870*. Boston: American Meteorological Society.

Manley, G. 1974. Central England temperatures, monthly means 1659–1973. *Quarterly Journal of the Royal Meteorological Society* 100:389–405.

Middleton, W.E.K. 1966. *A History of the Thermometer and Its Uses in Meteorology*. Baltimore: The Johns Hopkins University Press.

Millás, C.J. 1968. *Hurricanes of the Caribbean and Adjacent Regions, 1492–1800*. Miami: Academy of the Arts and Sciences of America.

Mitchell, J.M., Jr. 1953. On the causes of instrumentally observed secular temperature fluctuations. *Journal of Meteorology* 10:244–261.

Neumann, C.J., B.R. Jarvinen, C.J. McAdie, and J.D. Elms. 1999. *Tropical cyclones of the North Atlantic Ocean, 1871–1998*. Asheville, NC: National Climatic Data Center.

Nicholson, S. 1980. Saharan climates in historic times. In *The Sahara and the Nile*, eds. M.A.J. Williams and H. Faure. Rotterdam: A.A. Balkema, 173–200.

Ortleib, L. 2000. The documented historical record of El Niño events in Peru: An update of the Quinn record (sixteenth through nineteenth centuries). In *El Niño and the Southern Oscillation*, eds. H.F. Diaz and V. Markgraf. Cambridge: Cambridge University Press, 207–295.

Parker, D.E. 1990. Effects of changing exposure of thermometers at land stations. Chapter XVIII of *Observed Climatic Variations and Change. Contributions in Support of Section 7 of the 1990*

IPCC Scientific Assessment, ed. D.E. Parker. Geneva: Intergovernmental Panel on Climatic Change/World Meteorological Organization/United Nations Environmental Program.

Parker, D.E., C.K. Folland, and M. Jackson. 1995. Marine surface temperature: Observed variations and data requirements. *Climatic Change* 31:559–600.

Reid, W., 1838. *The Law of Storms*. London, John Weale.

Ropelewski, C.F., and M.S. Halpert, 1987. Quantifying Southern Oscillation-precipitation relationships. *Journal of Climate* 9:1043–1059.

Shapiro, L.J. 1982. Hurricane climatic fluctuations. Part II: Relation to large-scale circulation. *Monthly Weather Review* 110:1014–1023.

Smithsonian Institution, 1984. *Smithsonian meteorological tables*. Smithsonian Miscellaneous Collections, Vol. 114 (Whole Volume), Smithsonian Institution Press, Washington, DC.

Sparks, W.R. 1970. Current concepts of temperature measurement applicable to synoptic networks. *Meteorological Monographs* 11:247–254.

Tannehill, I. [1938] 1952. *Hurricanes*. 8th ed. Princeton: Princeton University Press.

Tukey, J.W. 1977. *Exploratory data analysis*. Reading, MA: Addison-Wesley.

U.S. Department of Commerce. 1998. *The Maury Collection. Global ship observations 1792–1910*. Version 1.0. U.S. Department of Commerce, NOAA/NCDC, Asheville, NC.

U.S. Navy. 1974. *Marine climatic atlas of the world, Volume 1 (Rev.), North Atlantic Ocean*. NAVAIR 50-IC-528. Washington DC: U.S. Government Printing Office.

Walsh, R., and A. Reading. 1991. Historical changes in tropical cyclone frequency within the Caribbean since 1500. In: *Wurzburger Geographische Arbeiten*, eds. R. Glaser and R. Walsh. Wurzburg: Mitteilunge der Geographischen Gesellschaft, 80:199–240.

Whetton, P.H., R.J. Allan, and I. Rutherford. 1996. Historical ENSO teleconnections in the Eastern Hemisphere: Comparison with the latest El Niño series of Quinn. *Climatic Change* 32:103–109.

Winter, A., H. Ishioroshi, T. Watanabe, T. Oba, and J. Christy. 2000. Caribbean sea surface temperatures: Two-to-three degrees cooler than present during the Little Ice Age. *Geophysical Research Letters* 27:3365–3368.

INDEX

in Caribbean, 6; rainfall amounts and timing affect crops, 6; use of slave labor to produce, 6
Sunshine index, 25, 62-4
Supplement to the Cornwall Chronicle. *See Cornwall Chronicle*
Supplement to the Royal Gazette. See Royal Gazette (Kingston)
Surinam, 15n

Teleconnections. *See* Climate
Temperatures, 1-3, 5, 10-11, 13, 15n, 18, 19n, 20, 42, 44-5, 51-5, 5-6, 77n, 78-80, 100-1, 103-4, 106-8; as measured by corals, 3, 55, 79; at Savanna-la-Mar, 52-5, 75; diurnal cycle of, 11, 52; earliest measurements at sea, 15n; global, 1; inter-hourly differences, 44-6; marine air, 51; sea surface temperature, 76-7; time of observation, 11. *See also* Thistlewood weather record; GOSTA; NMAT
Texas Gulf Coast, 130
Thames, 5
Thermohaline circulation, 100
Thermometer screen, 10, 44
Thermometers: exposure at Bread Nut Island, 43, 45, 108; Fahrenheit scale, 18, 107; Hauksbee scale, 18, 107; proper exposure of, 10-11, 107-8; types of, 10;
Thistlewood, Thomas, 7-8, 15-6, 18, 20, 22, 24-5, 28-9, 32, 35, 42-5, 47, 49, 51, 78, 81, 84-5, 88-9, 98, 101, 103, 106-7, 112, 114-5, 120, 122-4, 128-9, 131, 133, 135-8; birth, 16; changes in weather observing methods and procedures, 18, 28-9; correspondence with Edward Long, 16, 29; daily activity journal, 15-6, 128; defines cloud cover descriptors, 24; description of 1751 hurricane damage, 112-3; horticulturist, 7, 29; obituary, 7; manufactures own rain gauge, 29; perception of heat at different temperatures, 43; purchase of plantation, 16; quality of weather observations, 25, 32, 43-

4, 49; rainfall in Jamaica, opinion on, 8; renames Paradise Pen as Bread Nut Island, 16; sailor in East India Company, 16; 34 sails to Jamaica to find work, 7, 16; slave ownership, 7; suffered severe hurricane damage, 88, 127; surveyor, 7; survey work, 42; Thistlewood manuscript, 15, 15n; use of clouds and ships' sails to estimate wind, 40; visitors of, 7; voyage to India, 16, 34-5; weather observations compared with those of Thomas Jefferson and contemporaries , 8; 43-4; work skills and employment as overseer in Jamaica, 16. *See also* Thistlewood weather record
Thistlewood weather record, 7-8, 15-6, 18, 20, 22, 24-5, 28-9, 32, 35, 42-5, 47, 49, 51, 78, 81, 84-5, 88-9, 98, 101, 103, 106-7, 112, 114-5, 120, 122-4, 128-9, 131, 133, 135-8; abbreviates pluviometer as "pluv," 120n; begins to measure rain amount, 18; begins to regularly record wind direction, 16, 18; cloud cover and sunshine, 22-5; conversion of wind force terms, 39; database constructed from, 19-20; daily rainfall measurements began, 28; database construction and formatting, 19-20; discontinuities in descriptor usage, 28, 29, 62; elements recorded in, 16-7; encoded values of descriptors, 25-8, 109-10; estimated error of temperatures, 52, 55; his legacy, 8; inter-hourly temperature differences and statistical significance of, 44-5; land breeze, 37; meridional winds, 64, 78; miscellaneous phenomena, 40, 42, 110; no barometer used, 84; North (wind), 16, 32 110, 112, 134; observation time of temperature, 43; oldest weather record from Tropics, 15-6; rainfall amounts and rain days data, 55-62; rainfall and rain day descriptors, 25-32; rainfall minimum in, 75; scalar wind speed data, 40; sea breeze, 20, 37; sepa-